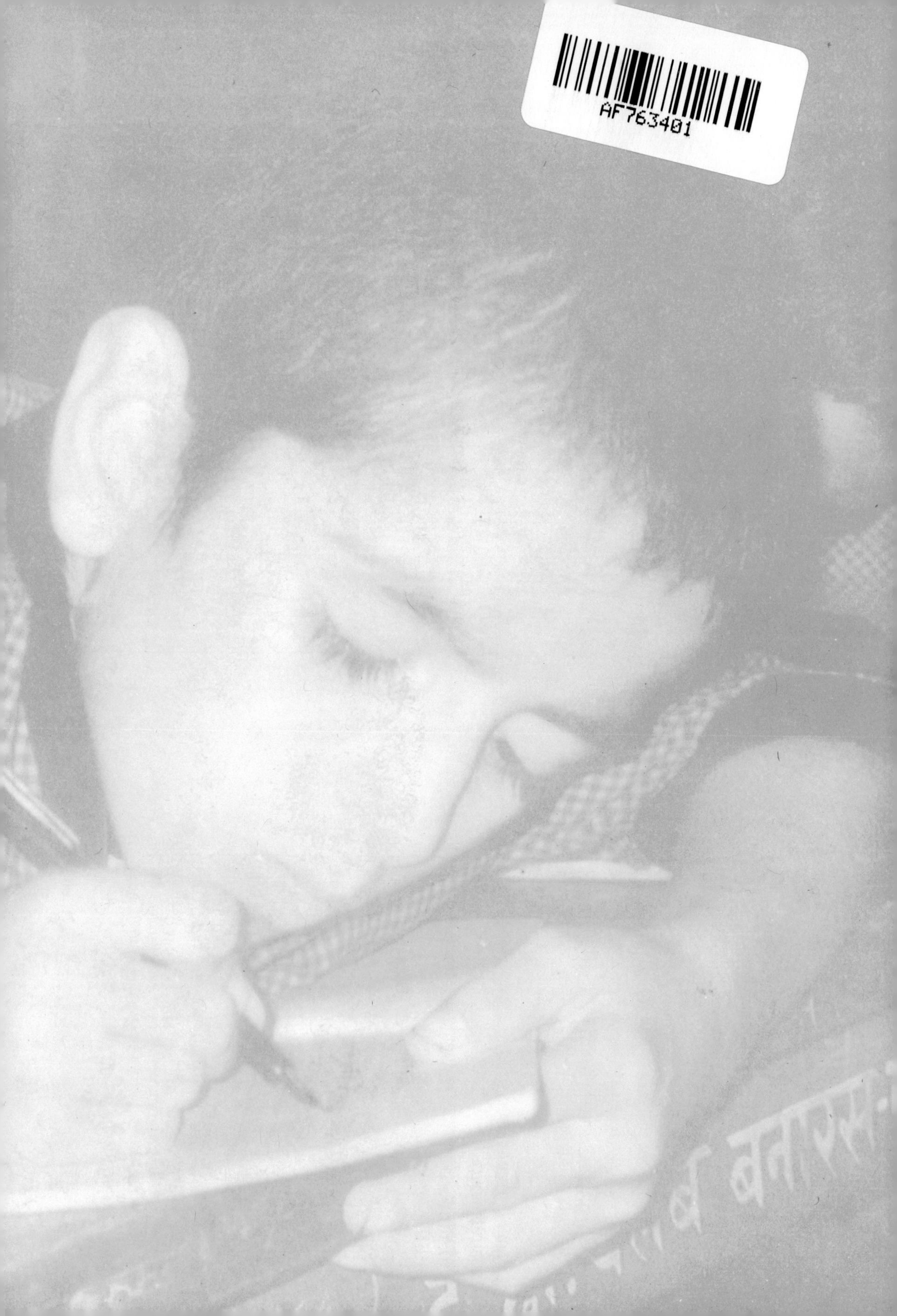

Culture and Development
No. 3

General Editor
Kapila Vatsyayan

The Cultural Dimension
of
Education

Unesco Chair in the field of Cultural Development

Indira Gandhi National Centre for the Arts

The Cultural Dimension of Education

Edited by
Baidyanath Saraswati

Indira Gandhi National Centre for the Arts
New Delhi

D. K. Printworld (P) Ltd.
New Delhi

Cataloging in Publication Data — DK

The cultural demension of education/
edited by Baidyanath Saraswati.
(Culture and development; no. 3).
Includes bibliographical references.

1. Education, Primary — Asia — Congresses.
2. Educational anthropology — Asia — Congresses.
3. Educational sociology — Asia — Congresses. I. Saraswati, Baidyanath, 1932- II. Series: Culture and development; no. 3.

ISBN 81-246-0101-1

First published in India, 1998

The views expressed in this volume are those of Participants, and are not necessarily those of the Editors, Publishers and Unesco.

Publishers
Indira Gandhi National Centre for the Arts
Central Vista Mess, Janapath, New Delhi - 110001

D.K. Printworld (P) Ltd.
Regd. office: '*Sri Kunj*', F-52, Bali Nagar, New Delhi - 110015

Printed at: DP's Impressive Impressions, New Delhi - 110059

Foreword

Kapila Vatsyayan

One of the major programmes of the Indira Gandhi National Centre for the Arts (IGNCA) is to launch multidisciplinary lifestyle studies of cohesive communities for evolving alternative models for the study of cultural phenomena and the inter-webbing of environmental, ecological, agricultural, socio-economic, cultural and political parameters.

In January 1995, a Unesco Chair in the field of Cultural Development was instituted at the Centre. As a part of the Chair's activities a four-day international conference on 'The Cultural Dimension of Education and Ecology' was held at the IGNCA from 13 to 16 October 1995. It called for a cross-cultural comparison and assessment of the problems involved in the fields of both education and ecology.

The proceedings of the conference are being published in two independent volumes. This volume deals with the question of education and development while the other (*Cultural Dimension of Ecology*) is devoted to critical issues pertaining to the natural environment.

Participants at the conference were men and women of experience and wisdom. They have been participants in the common endeavour of making education more relevant and meaningful. Chitra Naik in her prologue eloquently outlines the historical background of the state in which large sections of the globe find themselves. There are series of disjunctions and they manifest themselves in many spheres, most of all in education, specially formal education.

A system of education was established with the avowed goal of alienating the student from his immediate environment. Consequently, the moment of education became and continues to become the moment of uprooting the child from the culture to which he or she belongs. The numerous skills of literacy, numeracy and reason he acquires, the content of the education which is considered 'global' and 'universal', all make him or her an efficient tool in a vast machinery. The values inculcated are those of success, achievement, material progress of the little self in a competitive world. More, the marked emphasis on uniformity in a rigid system makes him or her an automaton. The driving force of his aspiration is immediate achievement and 'success'.

The few who go through the ladder of competition undoubtedly 'achieve', but in the process they are uprooted and certainly unaligned with the very ground from which they were nurtured. The larger number acquire minimal skills of literacy and bookish knowledge. Their harmonious world of work, function and ideation and faith is dead, and the new world is powerless to be born. Wastage is prevalent and unemployed educated youth people our lands. They are the human repositories of great energies which can be directed positively or negatively. The situation may differ in degree in different cultures of the erstwhile colonies, now young nation-states, but there is a similarity. With the exception of an infinitesimally small percentage who reach the pinnacle of the system and become global citizens, most others survive at the minimal level of comprehension, litle or no creativity and initiative.

In India, repeated Education Commissions were set up to reform and alter the educational system which was established by closing down *madarsas* and *pathashalas*, the *gurukul* and guild systems.

The reports of the Kothari Commission and the Education Commission of 1986 have called attention to the need for taking the cultural dimension of education into account. These reports recognised the need to reform the system in a manner that the world of work and the world of education, of home, family and education of individual and society, are not in conflict. Much earlier, Gandhiji had advocated a system of education known by its familiar names — 'basic education' and *nai talim.* Here manual and cerebral skills were in balance; the tools of education were through the use of the hands and the utilisation of local resources, natural and human. There was no undue emphasis on literacy.

After fifty years of experimentation, there is a renewed recognition of the relevance of the Gandhian model based on a total development of body, mind and soul, the values of restraint and self-reliance and both self-sacrifice and self-fulfilment through community participation. It is heartening to note that after a lapse of many decades and as a result of disillusionment with the present system of formal education, specially at the primary and secondary levels, many experiments have been conducted both in India and abroad. Mrs Oka's experiment and its success bear eloquent testimony to its efficacy. The Bose Foundation School is exemplary in its goals of achieving much in modesty. The Rangaprabhat experiment is unique in fostering the innate creativity of the economically disempowered. This is only to mention a few.

Now the question may well be asked: What have all these experiments to do with the cultural dimension of education, normally understood to be restricted to the arts? A reading of the papers by and sharing the experiences of Haku Shah, Dinanath Pathy and Nita Mathur makes it clear that the arts are a potent tool not only for acquiring artistic skills but more for attuning the child to his immediate environment. We will recall that Rabindranath Tagore had instituted Sriniketan in the proximity of Shantiniketan for this precise purpose. So also was the endeavour of Rukmini Devi in Kalakshetra and that of Anne Beasant. The J. Krishnamurti schools aim at a total flowering of the human personality, both within and without. The 'arts' no doubt are important and essential. The whole range of creativity through the traditional and age-old 'crafts' is even more important. The crafts are culture-specific; they transmit values which represent a symbosis of the function and the ideational, the abstract and the concrete. They can be individual activity, but are more often community activity. They are embedded in the rich oral traditions of this and other countries. This is the closest we can come to an understanding of incorporating the cultural dimension into education.

The plea in these papers from different points of view is to integrate the richness of this vast so-called unorganised and non-formal system into a new model of education or models of education which will not be deculturising, dehumanising and certainly not as wasteful. There is a great need for 'equity', for equalisation in terms of 'status' and value of the oral and the written, the manual and the cerebral, the local, regional and global. Equally important it is empower the culturally rich but socio-economically deprived, to be 'givers' as teachers and leaders, apart from those who have mastered the three R's.

Why has the intensity of this need increased in the decade of the nineties? The need for a more viable and productive system of education has become essential in a world which has experimented with a monolithic design of living. It has not brought peace and harmony. Tension, conflict, violence and intolerance are widespread. As obvious is the lack of ethical values and the indiscriminate use of the power of religion and faith to create wars and not peace. Where else except in education can new beginnings in creating human beings be made?

The child's education begins with the mother and then the father, and on to the teacher and from these to the community. Gandhiji had spoken of the expansion of oceanic waves. Education is the centre of these oceanic waves, the fragile fertile body, mind and soul of the child. If the process of education can attune him to himself and the world around, and if he can be skilful but not avaricious, knowledgeable but not arrogant, self-contained and not selfish, self-reliant and restrained by the curbing of greed, will we not have a better world cultrually and otherwise?

The conference was a modest attempt at articulating anguish, presenting a critique and showing the light of little lamps burning. However, it requires many thousands of such small and meaningful lamps of new education for a true transformation to take place.

It was evident that reforming the monolithic, large system was of no avail. Alternative models of significance, be it in education or cultural development, had to be instituted with a full recognition of the principles of plurality, multiplicity and yet inter-relatedness and interconnections. A network of coexistence. This is the spiders' web or fishermen's large net which contains all, but with flexibility and movement.

Prologue

Chitra Naik

I HAVE been listening with very great interest to the presentations that have been made, and I have found one common point in all of them: a deeply felt anxiety about what is not happening in education and a concern about what are the wrong things that are happening in education. It seems to me that this kind of discontent is essential if some action for change has to begin. In a way, therefore, this is a very good situation in which we should think about (a) how to bring about changes in education so that education gets related both to ecological problems and the culture of the people; (b) how education can become an instrument to ensure that man lives in peace with his environment and respects it for all that it has given; and (c) how man lives in real comradeship with other men, with other people.

These are very important matters that have to be culturally and ecologically examined so as to understand how we can evolve a faith in the basic philosophy of life which arose in India, in many Eastern countries and in many African cultures, that essentially there is no hostility, no adversary relationship, between man and man's environment. Actually, man is part and parcel of the environment, just one creature among the innumerable beings in the total spread of nature. Therefore, if man does not recognise the symbiotic value of the environment for surviving and growing, then he brings destruction not only on the environment but upon himself as well. This perception, which was there in the philosophy developed in these countries — sometimes consciously, sometimes not so consciously — resulted in the value system of man not being an adversary of the environment. This is something that we have to remind ourselves of, again and again, in the interest of man and his environment.

The first question to be asked is : What happened to these values? Why have they been weakened? The answer is to be found in the accidents of history. Our cultures, our

---

The views expressed in this paper are the personal views of the author and do not reflect or represent the views of the Planning Commission, Government of India, in which she is working.

countries, came in contact with certain other cultures, other nations, which had entirely different value systems, born out of the climatically inhospitable areas in which they lived. Take the West, for instance. Wherever we go in India, or even to the darkest places in other countries, something grows all the year round. You can at least eat some leaves, flowers, fish, or whatever is available. There is some food for most of the year. But the northern countries were often totally devoid of the means of day-to-day sustenance. In harsh winters, food became a problem. It was probably because of these hardships that they became marauders and explored the South. In their own countries they fought with nature and tried to conquer it, because nature was inhospitable to them. They were adversaries. When it snows so deeply that not even a blade of grass can grow, you can get angry with nature. Then you go somewhere else to search for food, become violent so that you may attack any creature that comes along, kill it, and eat it. There obviously was a geographical reason. The question is why people from certain areas of the earth became violent and why an adversary relationship got established between man and nature, and even between man.

On the other hand, there was a fundamentally friendly relationship between man and nature in other cultures; although there were aberrations too. Therefore, cultural geography may be a point to be remembered and studied. In recent years, some of our historians, archaeologists, linguists and Indologists have been investigating the historical origins of Indian culture from a fresh standpoint. Many researchers are engaged in finding out how the Indian culture originated. What was the initial value system, what happened later, where did the Aryans come from, who were they, did they come as adversaries and conquerors or was there nothing of this kind, and whether battles between Aryans and non-Aryans were the figments of the imagination of Western scholars who tried to reconstruct Indian history.

The pity of the whole matter is that rarely did Indian scholars make any serious effort to construct their own history with the necessary intellectual rigour. Maybe they were happy with whatever knowledge and social system they had and did not think it worthwhile to research their own origins. But now our own researchers and those Western researchers who are without the blinkers of superiority and are more empathetic towards the Eastern people, are inclined to think that as far back as 6500 years before the Christian Era, the *Rig-veda* had given an interesting picture of the local culture. This culture seems friendly to man's environment. Within that culture, everyone respected everyone and asked for cooperation and sharing. If you were *duhita*, a daughter whose duty it was to milk the cows, you took the milk to the Ganapati, who distributed it among the *gana*, i.e., everyone in the community. Community sharing was the custom and there was no competition. There was collaboration, cooperation, between human beings, between man and nature. Even today, some of our villages have contributory grain banks for emergencies. In some of the villages

I have been studying, people collect grain during the harvest, store it, and if any family is in difficulty — if a woman is widowed, if a child is orphaned — sustenance is provided by the community. These are parts of our cultural history and its continuation.

Studies and travelogues tell us that the cultures in the East, in India in particular, have known little violence. They have not generally been aggressive. One cannot ignore the instances where kings and others have fought battles among themselves to gain brides or treasures or 1000 horses or cows. But these were aberrations. Wanton destruction was not the rule of life. In the West, there seems to have been an endorsement of violence as between man and nature, man and man. The Western philosophy of the 'survival of the fittest' implies the wiping out of the weak by the strong, a philosophy which celebrates aggression. But in the East, it is the duty of the community, of the family, of kin, to look after the weak. That is the basis of oriental cultures all over. Another point in these cultures is renunciation, the shedding of power and authority with readiness. A novel by Pearl Buck shows that as soon as the mother-in-law turns 40, she hands over the keys of her domain to her daughter-in-law. It is another version of the *vanaprastha ashram* tradition that India has. We trust the younger generation to take over. Therefore, we go on educating it from the moment of birth, telling it, as it were, "Look, at some stage or the other, you will have to shoulder these responsibilities of mine: therefore learn the ways of life your forefathers have followed".

Enculturation as education becomes very important in the East. This informal education for cultural commitment is totally ignored now because of the onslaught of the West on our cultures. We do not educate our children ourselves. We leave it to schools designed by the West. How to overcome this cultural aggression called modern education is our real problem. Our anxiety about this is great and it is absolutely right.

In this context, let me take you back to Gandhiji and to the book called *The Beautiful Tree* by Dharmpal. Another very interesting study of indigenous education in India made by Joseph Di Bona, titled *One Teacher, One School*, relates to the same subject, i.e., basic education in pre-British India. Our administrators now decry one-teacher schools, although we need them in this country in remote hamlets. The kind of one-teacher school we had once upon a time, i.e., till the British school system arrived, has never been comparatively studied. There is an example of a one-teacher school in Di Bona's analytical account of indigenous schooling, where he has pointed out that in the indigenous school the teacher belongs to the community and the first rule was that he was not a bureaucrat appointed by an outside agency, a distant government. Why is it that now we hear complaints that the teachers do not go to teach? Obviously because they are aliens. They do not belong to the community, and they are not accountable to the community. They are accountable to a governmental structure and government rules and regulations which are distant from the people's aspirations and customs. This distant government does not know what the teachers are doing. Now, in this indigenous one-teacher school described by Di Bona, the teacher was a local person and was very greatly respected. The community gave whatever the teacher required. The children brought him vegetables. The weaver's child brought a *dhoti* or any other cloth he needed. The carpenter, the potter, the merchant — all gave something to the teacher. Some money was also collected for the school. Di Bona gives an example where the teacher received 32 rupees a month from the community. This was great wealth in those times, when you could get six maunds of wheat for one rupee. But what did the teacher do with this money? We are told that he kept some for himself but provided to the children writing materials and playthings, and soon even gave scholarships to poor children. The teacher was the manager of the finances of the school, and the community respected him because he kept very little for himself.

These teachers had a tenet from the *Upanishads* to guide them, to enable them to understand how man can become friendly with everyone. This tenet was '*gyan, seva* and *tyag*'. *Gyan* is knowledge. If you know the environment around you, if you fully know nature, then you become friends with nature, not an enemy. But how do you bring human beings together in a friendly relationship? It is through *seva*, through service. Service to others makes friends. And how do you get over the conflicts within your own mind, how do you prevent the battles within? By *tyag*, i.e., sacrifice. Thus knowledge, service and sacrifice formed the base on which the Indian teacher was expected to build up his work. We have instances of this in our educational history. But we have not been studying it properly. We have to go back to our own traditions, examine them, pick up the best in them and then mesh it with whatever else is desirable. Therefore, the best way for the merging of education, ecology and culture is to find and stabilise our value-base as indicated by the triple principle of '*gyan, seva* and *tyag*'.

For the pursuit of *gyan*, we must acquire scientific knowledge about our environment, about the universe. Scientific investigation of what is nature, what is the universe, what are we all, is the first base to reach. It leads you to the second base, namely, a philosophical as well as practical relationship between man and man. This can lead to understanding other people. This understanding leads to *seva*. It also leads to a very important principle, non-consumerism. The consumerism which has been brought into our Eastern cultures by the West has to be countered through *seva*. In the East, we were always inclined to be happy with limits to our wants and possessions. The Chinese mother-in-law, who handed over the keys at the age of 40 to her daughter-in-law, restricted her wants. She did not want personal power. People who adopted *vanaprastha* when they became old, shed power completely and handed it over to others. They did not desire anything after that except peace, contentment, time to meditate, time to look at their own lives, time to merge with nature. This was part and parcel of our culture.

The point is: What has happened to us? There has really been in the Eastern countries, and in African countries also, what you could call a destruction of our cultures. This happened because people came to us from entirely different cultures, cultures that tolerated and even glorified violence and greed. These cultures colonised us. Why could they do that so easily? Obviously we were a very simple people, welcoming and befriending everything that come to us. We are still very simple and 'colonisable'. Take the television onslaught. However stupid might be the shows on television, we do not protest. People have come to us from various countries. People came from as far as Mongolia to settle in Assam. People from Iran came to the western coast and have settled in this country as Parsis. People came from Abyssinia. One can find their descendants in the Konkan villages, called Habshis and Siddis by the local communities. People came to India from everywhere and were welcomed with open arms because our culture did not make us hostile to anyone. We accepted people, we absorbed them. We all became one Indian society.

In the same fashion, we welcomed with open arms the white man who came here. But we did not realise that the white man did not wish to stay with us, mingle with us. He had come with certain intentions which were not favourable to us. He wanted to trade, to exploit, to take away our valuables. He simply wanted the wherewithal to improve his own condition and to satisfy his greed. Therefore, whether these alien culture went to Asia, or Africa, or China, or to any other part of the globe, they plundered and suppressed the indigenous ways of thinking and living. We, in our simplicity, not only allowed them to plunder, but whenever they appreciated an ornament or other beautiful object, we generously gave it to them, feeling happy that they liked it. When one visits some of the huge mansions of the ex-viceroys and ex-governors of India, one can see the places filled with priceless Indian artefacts. Brass and stone images, jewellery, beautiful Banaras *sarees* and Maharashtra's *paithanis* used as curtains in their houses, all are evidence of plunder. But we cannot blame the plunderers. It is our fault that we did not make any effort to understand the difference between our cultures and theirs. When English was forced upon us as the medium of instruction, why did we not revolt? As a nation we are good linguists, we could have retained our languages as media and learnt English as well and some more foreign languages.

Some of our people naively thought that it was admirable to learn English and speak and write it better than an Englishman. Many of our scholars have done that. We thought that something new had come to us and we wished to master it. But gradually it mastered us, it mastered our minds. The cultural damage has been done. How is the damage to be controlled? Our schools and colleges have led to a complete splitting of our society into two.

This has happened in other countries also, for instance in Africa. We have a class of people that speaks English but is in its thinking partially English and partially Indian or African. There are culturally split personalities among those who have been educated in the British, French, Spanish, German, Portuguese and Dutch patterns of education. Unfortunately, these split personalities have been created among the best of our intellectuals, and many of them do not know whether they are Indians, Americans, Europeans, or whatever else. This cultural confusion seems to be the reason why we cannot readily sort out problems and make up our minds to take some firm direction. I am not talking only about the direction in education but in development matters as a whole. This cultural split seems to be the major handicap of the Third World countries which had been under colonial rule. Our elites have acquired much from outside; but whatever they had in their minds indigenously, culturally, does not go away. Therefore, we have a situation of mental conflict.

It is a very simple psychological finding that mental conflict obstructs decision and action. We cannot fully absorb the market culture, nor can we return to our own culture of sharing and sacrifice. We are like flying fish, neither fish nor fowl. Therefore, unless we cleanse our minds sufficiently from the cultural impact from outside, we will not be able to think clearly. This does not mean that we should not study, appreciate and mix with other cultures. We must seriously study them. At the same time, we must study our own cultures scientifically, dispassionately, in order to understand ourselves. The perceptions that others have of us must be allowed, but we must similarly study other cultures. Our scholars, our researchers, need to study the cultures of Europe and the US with real scientific rigour. It would help us see ourselves in a broader perspective and promote indigenous scholarship. The violence and misery which mark many a culture today must be investigated by us as a third party. Our scholars need to do this. It is not healthy that we eat up ourselves and feed on ourselves for our research effort. Globally gathered intellectual food should be served by us for ourselves. If we start doing that, the question of education, culture, ecology, and everything that matters in life will start becoming clearer to us, because we will then be ourselves, which today we are not.

The question of medium of instruction is raised again and again in our country. There are numerous countries which use their own languages as media of instruction right up to the highest level of learning. Only certain ex-colonised still use the languages of the colonisers because they have lost their self-confidence and have been overwhelmed by the scholarship couched in that particular language. No doubt, such scholars as Wheeler, Max-Müller, William Jones made seminal and significant contributions to the study of archaeology, Sanskrit literature, Indian languages and so on. But we must ask why we did not build such critical scholarship ourselves. We knew our religions, languages, flora and fauna, sculpture, architecture, arts and crafts, politics, economics and all that makes up a cultural fabric. We are now told that Panini gave us a grammar which can be used for computerisation and that Sanskrit could be the best language for global communication. But that grammar was brought to light by Western scholarship. Of course, there is now a glimmering of realisation that we have neglected to study ourselves. A state of discontent has begun to take form on this score among our intellectuals. This is a good sign. We have started searching for our own cultures and their meaning for us in modern times. This search must be intensified. If the new awakening is to be sustained and enhanced, our educational perceptions must change.

Education has to be investigation, autonomous self-learning in collaboration with teachers and co-learners. This principle must be reflected in pre-schools, primary schools, and all along the line. We must also ask, what is a school? Is it only an arm of the bureaucracy? If so, how can it be changed into a stimulant of investigation, of self-learning, of change for the better? The people do not know what goes on inside the schools, particularly in the rural and tribal communities. They do not know what the curriculum is. The textbook has often become an instrument of conditioning the minds of people. Wherever an educational system is totally managed by government, it inevitably pressurises teachers, students, parents and the public to behave in a particular way, which is decided without consulting them. Although our educational system is not altogether totalitarian, the governmental concern for people's participation is hardly evident in actual practice. In the rural areas, participation is construed as asking the villagers to give something or other to the schools. A School Improvement Programme is mounted by officers of the education departments and schoolrooms are constructed or equipment is collected by asking the villagers to give free labour or donate equipment or money to the school. Then the poor people do collect money and articles to be donated to the schools. But whose schools are these? Who owns them? Government owns them but does not take full responsibility to make them function properly. It then tells the people to please participate, help government, invest in the schools, although we own them and you do not.

What can this be called? Participation of the people or their exploitation? This is certainly not people's mobilisation, because mobilisation is the means of empowerment. And what do we mean by empowerment of the people? The people are already culturally empowered since, over generations, they have learnt how to survive and how to look at life philosophically. But they are not empowered in managing new knowledge. They are powerless in the sense that they do not know what we know and with which we have improved our material condition. Of course, we too do not know what they know. This creates a cultural divide. How to build bridges over these chasms between the educated elite and the people, how to make education a concern of the people, how to enable them to own their schools? What would be the strategies of such planning?

Today there is a trend to send our scholars and officials abroad to learn the techniques of educational planning. Officials and researchers who are already removed from the people work out demographic data and certain educational systems they might prefer. How to gear educational planning for building bridges between the elite and the masses is our real challenge. In these exercises people are reduced to statistics. Then planners allocate funds to these statistical tables and not to the variety of people who require education of various kinds. The planners overlook the fact that not only funds but the people's intelligence is a resource. Their capacity to solve their life's problems, despite poverty, is a great resource. The older community members are a resource for looking after children and the women in each family are a resource for keeping the family properly managed within a small income. Our planners do not remember these resources since certain macro indicators mean much more to them and foreign economic models attract them more than the inherent resources of the people whose education they propose to plan. Therefore planning goes wrong.

If we are to do some meaningful planning, two things are necessary: first, we must educate ourselves afresh in order to understand our people and their cultural resources; second, we must honestly collaborate with the people in planning development as jointly perceived by them and by us. For instance, environmental destruction can be halted if we understand that the people's cultural consciousness respects trees, water, flora and fauna, the hills and the valleys. They live with these but the exploitation comes from outside, from the commercial interests of the wider world which do not relate to the local cultures and their environments. As a result, the people are often made to do what they really do not wish to do. They do not wish to exhaust water resources and pollute the air. The money economy, which operates unknown to them and over which they have no control, upsets their culture, their livelihood. Neighbourliness disappears when the barter system is replaced by monetary transactions. However, some bartering of services, and of the wherewithal to live, does exist to some extent in our villages even now. This requires protection against the pressures of the money economy in all aspects of rural life.

This culture of sharing of services, of goods, has been a valuable parcel which children have learnt at home. This is excellent education. But when they go to school, there is no sharing. Every child must have an individual school bag, slate, books; a confidential examination is given individually, and so on and so forth. There is no sharing in the schools of the government although there is sharing in the children's culture. As a consequence, the children get culturally confused. They do not understand why the school culture is different from community culture. We urgently need to look into this matter. Our education must be so redesigned as to prevent this splitting of personalities because of the clash of cultures.

It is necessary to welcome science and technology; but it does not mean the suppression of the innate conviviality of human beings. Science is universal and there is no science which belongs to any particular country. But while science is universal, cultures are specific to various communities in the world, based on their history, geography, and the resultant viewpoints on life, on human relationships and human purposes. Science and technology have to be accommodated within a given culture at the discretion of the people in general and not at the behest of a culturally alienated elite. Such adaptation of science and culture in a relevant fashion has to be woven into a strong social movement.

People have been talking of political will and of the government doing certain things. But governments are a part and parcel of society, and therefore it is society that must give a direction to government. It is the culture of the people that must assert itself in combination with the scientific, future-oriented outlook. It is necessary to arouse the social will to study our problems of education and ecology. If we can mobilise the people to building up a social will for change, or, in other words, for cultural adaptation, for a fusion of the old and new and not a weak compromise, we may find answers to our present educational and ecological predicaments. What we need is a clear philosophy, a value system which helps us (a) acquire such knowledge as might end the conflict between ourselves and nature, (b) strive to end the conflict between man and man, (c) emphasise sharing so as to remain free from greed, and (d) even to sacrifice so as to end the conflicts within our minds. If such a philosophy and concretised value system could be infused in the process of education, and if the people could 'own' the system to protect its cultural relevance, our educational and ecological problems could be properly dealt with. What we need is educational action on this basis, starting with mass education, primary education. Even small plans of innovative action all over the country could help us. We need not visualise, at this stage, any comprehensive national, state-level or district-level action. If we can have several experiments, alternatives, which come up with people-generated ideas and action from place to place, whether in India or Africa, or any country of the Old World, we would be able to put together our fragmented cultures and overcome the pressures from cultures totally alien to our own so as to prevent further fragmentation.

Contents

Introduction

Baidyanath Saraswati

WE, the people of the 'developing' countries, are struck by the contrast between the present age of 'modernity' and the wisdom tradition of a past time. Yes, there are today extraordinary technology, ultra-rapid communications, remarkable information networking, and exciting and challenging changes in science itself. But where is the Man — the enlightened Buddha, the compassionate Christ, the loving Muhammad, the truthful Gandhi? With all the wonders of modern science and technology, where are we going? Are we not moving into the New Age with the burden of a new myth of materialism? Are we not suffering from the disease of machine mindedness? Most of us are asking such questions. Some say that these advances cannot bring into being a normal civilization. Others think that there is a profound way: cultivating wisdom from all the knowledge that mankind has gathered. Now, can we take on the responsibilities called for by the future of humanity?

In January 1995, a Unesco Chair in the field of cultural development was instituted at the Indira Gandhi National Centre for the Arts. As a part of the Chair's activities, four field studies were carried out on the trends and problems of primary education and the natural environment. A Conference on the Cultural Dimension of Education and Ecology was held in New Delhi on 13-16 October 1995 to pursue exploration in these fields. The sixteen essays collected here refer to the state of primary education in Bangladesh, India, Indonesia and Thailand. In particular, the experts have called attention to (a) **evolving** new perspectives in primary education; (b) **developing** Man the artist; (c) **promoting** a genuine culture of peace; (d) **re-considering** Gandhi's basic education; and (e) **re-examining** national policies on education.

Evolving a New Perspective

All cultures do not share the same vision of life. The spirit of a 'traditional' culture, for instance, is not the spirit of 'modern' culture. This is clearly reflected in their conceptions and practice of education.

- Traditional education aims at expanding the spheres of existence by **social awareness** (forming kinship with the entire world), **cosmological awareness** (expanding of being by self-transformation) and **technological awareness** (relating creativity to the ritual enforcement of life).

- Modern education, in contrast, **teaches** a way of life limited by self-centred consumerism, **allows** man's ego to establish itself as the conqueror of nature, and fragments people through competitive vocations and specialized technical professions.

How do we resolve this deep dilemma between traditional and modern systems of education? The authors of these essays provide a renewed sense of awareness. In most experiments, what is being evolved is the 'Middle Path': without one extreme, without two extremes.

Gedong Bagoes Oka, an Indonesian follower of Gandhi, is in sympathy with the holistic approach, a Gandhian experiment, which

- draws from the noble Vedantic dictum *tat twam asi* the oneness of life;

- aims to bring out and foster all the potentials in the child and help it express these through *ahimsic* (non-violent) channels; and

- makes the school a happy adventure of discovery for the child.

Dwarko Sundarani, a Sarvodaya social worker based in Bodha Gaya, also stresses on

- total development, i.e., physical, emotional, intellectual and spiritual;

- the correlation of education and manual work, leading to self-reliance for the basic necessities of life; and

- a harmonious relationship with nature and society.

Shakuntala Bapat and Suman Karandikar, with research experience in the rural areas of Maharashtra, favour

- indigenous basic education consisting of elements which are local and culture-friendly.

Baidyanath Saraswati, Shivashankar Dube and Ram Lakhan Maurya, who experiment with self-organizing educational systems in the ancient city of Varanasi, trust that a humanistic appreciation of tradition with a universal significance can be develoved through

- *swaraj* (self-rule), *swadeshi* (homemade), and *sarvodaya* (enlightement for all) in and through primary education, without reference to government grants, with limited resources in men and money.

The perspective extensively introduced in these experiments, one would hardly deny, is 'new' in relation to the present enterprise of the modern system of education. It is not a new creation of thought or method: its newness lies in the restoration of a self-grounding tradition.

Developing Man the Artist

Modern cultures transform more readily and more rapidly than traditional cultures through 'deterministic development', which means making the technologically designed environmental schemes and politically governed social organizations a more or less workable fit. Traditional cultures, on the other hand, blossom by the sacred art, which always

- implies transformation, a real understanding of aesthetic intuition;
- captures the spirit of the invisible order of life (creativity);
- illumines the organization of thought in terms of types of activity, and
- advances in the realization of humanness.

How relevant is art in the age of science? To understand the relevance of art in every age, one has to perceive all of existence in the unity of thought and action, because as a method of gaining self-knowledge, art is an intellectual pursuit and as a creative work, it offers an aesthetic experience of beauty and goodness.

Haku Shah, a distinguished artist and a follower of Gandhi, gives art a natural place in education and points out that

- the child has an innate capacity to express itself through available material and things, even the sand on the beach or in a desert.

Truly in the spirit of Coomaraswamy (who held the view that an artist is not a special kind of human being but every man is a special kind of artist, or else he is less than a human being), Haku Bhai cites the example.

- An illiterate singer, his wife, and their children started painting the most unusual themes like themselves instantly upon a mere word of inspiration uttered by him.

D. Patnaik, another great artist and a good Gandhian, whose experiments in paintings and plays are widely recognized, narrates his experience.

- Students of drawing and painting enjoy learning crafts such as carpentry, sewing, book-binding and image-making.
- Ideas of character building in children can be generated by staging plays and writing poems.
- A good teacher is always ready to take the risk of guiding elitist, mindless and directionless students.

Dinanath Pathy, a teacher of visual art, is of the view that

- interlinkage between culture, education and ecology is the very essence of human life;
- art in school should form a part of the total learning system to provide an aesthetic orientation to the child, whether it is in mathematics or in science, geography or literature;
- art is not the negation of science, technology, and modern living; and
- art is a rejuvenating tool.

Ravi Chopra, a young social worker who runs a photographic training institute for street and slum children, takes the position that

- photography enhances the artistic bent of the children's minds,
- increases technical ability,
- keeps children occupied with an avocation, and
- contributes to their creative expression, self-confidence and employment potential.

N. Radhakrishnan, an eminent Gandhian scholar who recognizes and 'sees' the light in theatre-in-education, calls attention to the fact that it

- helps children sustain interest in schooling;
- develops self-sufficiency and leadership qualities; and
- enables them imbibe values and attitudes which the formal education set-up does not offer.

Nita Mathur, a young student of anthropology who studied the transmission of the Bharatanatyam form of dance among the Tamil Brahmanas, arrives at the conclusion that the dance

- unveils to children many channels through which emotions and states of mind are expressed;
- leads to the cultivation of perseverance, reverence, and tolerance; and
- develops a holistic life-style and perspective.

The artist-authors of these essays, deep in their experience and imagination, do not mention a definable methodology of art. Instead, they talk of character building, human dignity, and discovery of one's own self. The introduction of dance into the lives of Tamil Brahmanas at the pre-natal stage is a case in point concerning the view that 'Man is a special kind of artist'. It also tells us that the transformation process begins in the womb in a natural way.

Promoting a Culture of Peace

The present age of modernity carries within itself an internal necessity called 'development'. Tradition is viewed as 'anti-development'. The discourse on tradition versus modernity has one point of agreement: education as a factor in peace.

Pataraporn Sirikanchana, a Buddhist philosopher and historian of religion, highlights the contribution of developer monks to the Thai culture of peace by way of

- making young people and children familiar with Buddhism;
- inculcating moral discipline and cultural appreciation;
- teaching young people and children a way of life that follows Buddhist principles; and
- training young people and children to work for public welfare.

Experiments at the Bose Foundation School (Saraswati *et al.*) have shown that there is an echo of positive change in the multi-religious environment of non-literate communities, to the extent that

- Muslim girls fettered by convention to profit by new educational systems now stand first in all the seven classes and first in the school.
- Their customs forbade them in the strongest manner never to go outside

their houses wihout the veil, but they now come to the school unveiled and excel in public performances in the arts.

- Children of this school have learnt to restrain communal passion and their guardians, especially mothers, have realized the value of schooling.
- Those who had rejected the Foundation's vision of cross-cultural education now see a culture of higher things evolving.
- Those who once used to throw stones at the Foundation School are now the most frequent spectators at their children's performances.
- Finally, in their direct experience of working, the organizers and teachers find the flowering of their service, the feeling of heart, the hope of resilience, the harmony of the rhythms of life and knowledge, and a noble culture of peace evolving.

Nita Mathur makes a reference to the contribution of art to the culture of peace, particularly in that

- dance performs the cathartic function of releasing pent-up emotions and drives.

War and peace are both real. Quite evidently it depends on what one wills: war or peace. Hence the importance of producing the actors of peace. Peace through a spiritual vision of shared responsibility is more enduring than the political economics of peace. The educational system has its own unique process of establishing peace and concord.

Reconsidering Gandhi's Basic Education

Emerging from the European colonial context and influenced by techno-political expediency, there has been a great ferment in educational thought in India. Many reject the colonial form of education offered to them. M.K. Gandhi and Rabindranath Tagore gave us an alternative system. Both emphasized the education of the self through creative activity.

Devi Prasad, a famous ceramic artist who had joined Gandhi's *nayee talim* (new education) or basic education, characterizes the new system in terms of

- mother tongue, or the language of the area, as the medium of instruction;
- top priority given to society's relationship with Nature;
- emphasis on creative activity as a vocation for livelihood and a way of

gaining knowledge;

- drawing out of the best in child and man;
- education being self-supporting;
- education covering the whole of life from the moment of conception to the moment of death;
- education as the only method for the true development of the mind;
- making cleanliness a science as well as an art;
- equality of all religions; and
- self-evaluation, along with evaluation by teachers, as a method of judging the all-round progress of a student.

Parthasarathi Banerjee, a scientist, comes out with a theoretical formulation.

- Basic education is not primary.
- It aims at skill formation, attitude development, conducting a good practical life, and imparting a *vidya* (virtue of the good).
- It can cater to any age group or to anyone having the desire to conduct a practical and pragmatic life.
- The Indian practice of basic education puts great emphasis on a source of injunction (the guru), an injunction (on what not to take), a causal description in the form of a theory, a set of names of the terms for which an action has been undertaken, and a set of rule-based integrated practices or action-orientation.
- The distinctive features of the institution of basic education are that : (a) it does not segment education in a hierarchy but divides knowledge and discourse epistemologically and therefore non-hierarchically; (b) it is not based on a divide between theory and practice and does not reduce science to a set of items of information; (c) it brings knowledge to the common people; and (d) science no longer remains in isolation but is created anew skilfully, while the discourse with abstract theories of science continues through narration, description, and an attitude of truth.

Here the artist and the scientist have exhibited with an equivalent intensity of conviction that

Gandhi's scheme of 'basic education' is a dynamic concept which has the potential for the total development of man. Undoubtedly, this scheme is the most epoch-making event in the history of education in modern India. Its uniqueness lies in its

- comprehending a development process, the true meaning of education, not limited by age, not graded in terms of 'higher' and 'lower'; and in its
- considering the whole period of human existence, hence — education as culture.

Re-examining National Policy on Education

With the end of British rule on 15 August, 1947 began the preparation of plans for a comprehensive educational reconstruction in India. The idea of planning and administering education nationally is a 20th-century concept. The post-Independence national policy on education is understandably in a state of flux. The picture is unclear, but the narrow national concern is evident in India's national policy, which aims at

- promoting national progress;
- creating a sense of common citizenship and culture;
- strengthening national integration; and
- giving greater attention to Western science and technology.

K. Gopalan, a retired education officer of the Government of India, narrates India's strategies to achieve the universalization of elementary education (UEE), which has three aspects, namely, universal access and enrolment, universal retention up to 14 years of age, and a substantial improvement in the quality of education. Admitting that UEE is still an elusive goal, he highlights some of its major initiatives and strategies that have led to

- disaggregated target setting and decentralized microplanning;
- strengthening alternative channels of schooling such as non-formal education;
- introducing minimal levels of learning;
- improving school facilities by revamping the scheme of Operation Blackboard;
- establishing linkages between programmes of early childhood care and education, primary education, literacy, and UEE;

- addressing the more difficult aspects of access, particularly to girls, disadvantaged groups and out-of-school children;
- restructuring teacher training in view of the changed strategies and programmes;
- availing of external financial support for basic education; and
- launching the National Elementary Education Mission.

Neeru Nanda, an educational administrator in the Government of India, writes a critique of the prevailing school system which shows that

- free India has no will to fulfil Mahatma Gandhi's dream of reviving the ancient tradition of the village;
- instead, the government educates the masses through a vast centralized machinery and superstructure of staff, infrastructure and resources;
- universalization of primary education, the adult literacy campaign, and the investment thrust on primary education to produce the **literate child**, carry within themselves the logic that the goal is unattainable;
- in the public perception the system of mass government-sponsored education appears to have failed to deliver the goods, increasing expansion having led apparently to decreasing quality;
- by initiating the expensive Navodaya Vidyalaya Scheme for rural children, the government simply brought the process of handling contradictions inherent within the systems to its logical climax; and
- the search for alternatives by the government and non-governmental organizations suffers from all the general weaknesses of the state system.

A.M. Sharafuddin, an education officer in the Government of Bangladesh, gives an account of the many innovative programmes in primary education launched both by government and by non-governmental agencies, whose major characteristics are that they should

- be based on direct experience and practice;
- be cost effective;
- show easily verifiable results;
- have interdependence within subsystems;

- have in-built mechanisms for measuring achievement;
- have simple but effective management; and
- create a strong team spirit among their members.

On the role of non-governmental organizations introducing non-formal education programmes in Bangladesh, it has been pointed out that there are some obvious limitations such as the issue of replicability of the programme and the lack of resource materials. The overall impression of the Bangladesh NGOs is that they have been successful in their mission. NGOs in India are generally in favour of hasty expansion. So is the Government of India. Should **quality** have a prior claim over **quantity**, the national policy on education will have to follow a different path.

The Unanswered Questions

- Must we make the nation above and beyond culture?
- Must a nation model the minds of its people?
- Are people bound to a nation by an unending chain of dependencies?
- Are national policies on education a 'new' beginnings (restoration, imitation, creation)?
- Must we allow the individual man to be fragmented by opposing forces, powers and politics?
- Must the intellectual universe be brought to a sterile uniformity?
- Can a single system of education serve best for thinking and for knowing and experiencing the multilayered and multidimensional reality?
- Can the orderliness of a culture come from the language (mother of reality) of another culture?
- Is 'modern' education emancipatory or corrupt?
- Is the written word the real 'human', 'social' world?
- To what extent is man the creator of knowledge?
- To what extent is technology the paradigm of human life?

1

Gandhian Experiment in Primary Education

The Story of Taman Kanak-kanak 'Gandhi'

Gedong Bagoes Oka

Nothing turns out right so long as there is no harmony between body, mind, and soul.

— M.K. Gandhi

'TAMAN kanak-kanak' is the Indonesian term for kindergarten, and in further references the letters TK will be used.

As background to TK 'Gandhi' in Candi Dasa should be mentioned the Bali Canti Sena Foundation, established in Denpasar, Bali, on 10 December 1970. Its objective was to meet the needs of the time, i.e., to disseminate Gandhian ideas which would give the right orientation to our young generation and make them responsible citizens of the Republic of Indonesia. This, we believed, could only be done when the basis of our life and attitude was *ahimsa* and *satya*, the great twin principles of *Vedanta* successfully demonstrated by Mahatma Gandhi. We endeavoured to do this by starting a monthly that would bring out suitable articles to promote open minds, a spirit of enquiry and constructive activities. Holding camps and outdoor activities for students were part and parcel of this endeavour.

Several of such engagements, however, met with little success. It turned out that camps and outdoor activities were popular, but little was achieved in terms of sustained work. The Gandhian way required a more solid interaction.

Then with great trepidation we made bold to try out a Gandhian Ashram which we named 'Canti Dasa'—servants of peace—which, as if so ordained, was located at a beautiful spot at the foot of an ancient temple.

The urge to be of some service to our immediate community led to a concern for health. A modest dispensary was started, which brought in its wake a free lunch programme for our neighbour, the village school with its 300 or so pupils.

Hardly a year had this project been underway when we decided to suspend our activities due to strained relations with the provincial religious body, who were suspicious of the motives behind the lunch programme. In order to continue our feeding programme, we hit upon the idea of starting our own school. Our financial limitations compelled us to be modest and imposed severe economy in our approach. For this purpose a kindergarten seemed to be feasible, since our area had no kindergarten.

Candi Dasa derives its name from an ancient temple built around the year 800. It must have been some kind of retreat or hermitage, judging from the austere architectural style. It is an exceedingly beautiful spot: a strip of golden beach with hills as a backdrop, on top of which is perched this austere temple. A small fresh-water lake fed by many springs on its banks runs into the ocean.

On the banks of the little lake is Ashram 'Canti Dasa', dedicated to the great soul of Mahatma Gandhi, whom we look upon as our guru, although he himself during his life emphatically refused to be anybody's guru. Such was his humility and honesty.

Looking back over the years we realize that, unknowingly, we started some pioneering work of the most difficult kind, i.e., to change people's minds by using persuasion and personal example, the Gandhian way *par excellence*. And that, it should be added, in the teeth of opposition of vested interests.

With a kindergarten to be developed, our own objectives and goals began to take clearer shape. As a concrete manifestation of the Bali Canti Sena Foundation's aim, which has for its guiding principles *ahimsa* and *satya*, the objectives of TK 'Gandhi' are:

(a) to apply a holistic approach to the education of our pupils. The holistic approach, which is derived from the noble Vedantic dictum *tat tvam asi* — oneness of life — should be reflected in the methods and methodology of the school;

(b) to bring out and foster all the potential in the child and help it express these through *ahimsic* channels; and

(c) to make school a happy adventure of discovery for the child.

Candi Dasa cannot be properly called a village, as it consists of a score of hamlets sprawling across the beach area of the village of Bugbug. Its inhabitants number not more than 750 to 800 people. Their source of living is petty farming. Only coconut groves thrive in the locality. Cattle raising adds to their income, but in the dry season grass is scarce. Fishing is also taken up, but it seems that of late the catch has decreased due to superior methods of fishing by people with capital. Of the lime kilns that used to help a few families, some have had to be closed down due to the enforcement of laws supporting ecological balance. Even from colonial times this area has been a poor one and many of its people have over the years migrated to more prosperous parts of the island and of late also to other islands.

There used to be one village school only, our neighbour, but the new drive for education made Candi Dasa one school richer. However, now there is already the problem of filling the first grades of these schools with enough pupils.

Apart from two or three carpenters and one blacksmith, no village industry can be discerned. These are only to be found in the adjacent village of Tenganan, which is famous because of its indigenous government and life-style that has remained intact for almost a hundred years. It has become one of the highlights of tourist interest. Although the tourist industry tries hard to sell this area it does not seem likely that hordes of tourist will come in the near future.

Experiment

In Indonesia even private schools have to follow the syllabus set by the government, with some concessions so long as there is no clash with the country's philosophy, *Panchashila.* In practice, thus, there is some leeway for teaching to depart from the government's dictates. In fact a good deal depends on the creativity and inventiveness of the teacher, who should know how to interpret the spirit of directives creatively. The teachers should be resourceful in integrating and weaving their own outlook and approach towards education into the government guidelines.

The ultimate objective of our educational approach being the attainment of true freedom of the individual, the immediate goal is how to foster those qualities and potential in the child that would help to bring about a peaceful, non-violent human community where relationships are conducive to harmony and creativity.

The method of teaching emphasizes the self-activity of the child, in which the teacher participates and stimulates rather than teaches as understood generally. The main idea that a teacher should bear in mind is that whatever is done in class should be rooted in the children's environment. Whatever is artificial should be avoided; in other words, there should be harmony between life at home and in school.

The poverty prevailing in this area means that parents should be saved school expenses. Therefore the only money parents spend is on the few things absolutely needed in class, such as two or three cheap copy-books and a pencil every other month, while no fees or other expenses are charged.

It goes without saying that in our system the principle of *swadeshi* is given prime importance. To make or produce whatever we need for our food or use is religiously practised in our school. This is possible because we grow our own food, have our own carpenter and tailor/dressmaker as well as musicians. The material we need also should be found in our environment.

Teachers

Ashram Canti Dasa, which runs the TK 'Gandhi', is truly a self-supporting community and receives no aid or subsidy from the government or other sources. Therefore our financial means are very limited and only the utmost economy enables us to survive. In such a situation there is no question of hiring teachers. When we opened the TK 'Gandhi' qualified kindergarten teachers were scarce anyway, while those who had been trained in government institutions were cut according to government taste. Luckily a teacher was found who due to political stigma had been jobless for a number of years. She was more than willing to accept our offer, even on a truly 'honorary' basis, her salary being of 'a basic need sufficiency'. The good luck was that she was a good singer and could teach the basics of Balinese dance as well.

After two years we drew the central government's attention and were then given by way of support a woman teacher to help out in our TK. However, as is usually the case with government employed people, she proved more of a liability than a real help. In effect all Ashram members get turns in assisting in the TK whenever they are free from their own duties in the Ashram. For everyone of us knows something that could be put to good use in the TK class. One older woman is a master at making offerings and palm frond cuttings, so indispensable in Balinese ceremonies and decoration. Another knows how to play the flute and yet another girl teaches swimming. Thus our TK has quite a few different teachers in a week. This is how we make the most of our limitations.

Finding the children to people our newly opened TK also presented a slight problem at first. As is well-known, in farming families children have to help out at home or in the field at an early age. For girls, from the age of four daily task could be from collecting firewood, minding younger siblings or just watching the new paddy harvest drying in the sun. Little boys are made to mind grazing cows or the ducks that are let loose in fields that have just been harvested, or perhaps simply watch the house while their mothers are fetching water.

With the cooperation of the village chief some kids were rounded up and within a month we had a class of 27 children. We considered ourselves fortunate that someone, and a woman at that, could be found willing to mind the children. To disseminate the ideas we nurse about education through this TK, all the members of the Ashram eventually became some sort of teachers according to their skills.

Our children come from very poor backgrounds. Their parents struggle the whole day just to scrape a living. Therefore the principle of *swadeshi*, aside from being an ideal — one that we try to follow faithfully, is pure necessity.

Teaching aids that seem to be a common sight in city or town kindergartens are conspicuous by their absence here. Thus, in this regard, the recourse taken to environment is a must. A real blessing for TK 'Gandhi' is its paradisiacal environment. We are, as it were, drenched in the gifts of nature: water, air, both pure and clear, plants in all shades of green, and healthy-looking animals and pets, the expanse of the blue ocean take our children daily in their happy embrace, stimulating in them free spontaneous movements, laughter and song. That our place is a veritable paradise is apparent from the fact that most of the children are already on the TK premises long before school starts. We can see them on the swing or the see-saw, running back and forth on the sandy beach, the little girls walking hand-in-hand while collecting edible berries and fruit that has dropped to the ground; or intently watching a newborn calf drink milk from its mother.

With the hills right behind us, as well as a fresh-water lake, the Ashram's site indeed fulfils the three requirements mentioned in the *Vedas* to make it a fit spot for meditation and other noble pursuits.

> Where the hills meet the ocean
> And there are springs nearby.
> Verily, that is the place for meditation.

The surrounding coconut groves offer plenty of yellow palm fronds for the children to practise on when weaving little mats to sit on, or when cutting decorative streamers and a variety of geometrical forms so indispensable in our celebrations.

The beautiful beach at the doorstep of our school can at any time be used as a vast sandbox in which the children can model, build castles of sand or simply draw and scribble to their hearts' content. Once the government supervisor of kindergartens came to inspect our school. His only remark was, 'The only pity is that there is no sandbox here for the children', to which our teacher replied: 'Sir, the whole ocean and its beach is the best sandbox one can think of'. What an apt remark to show how hidebound we often are, a result of our parrot-like education.

Twice a week the school goes for a walk in the nearby cluster of hamlets or in the hills, for the beach is their daily fare. This is the occasion to make them aware of the beauty of our surroundings and the industry of the villagers. For at any given time some activity is going on, be it feeding the pigs, pounding rice, weaving mats or a carpenter giving shape to the boat under his hands. Then all their senses can be stimulated, which is later reflected in the drawing class.

Swimming is almost a daily feature of our curriculum. Simple exercises and postures, derived from yoga and Balinese dance, are integrated in their physical exercise hour, while indigenous children's games are promoted.

We do not follow the teaching of the three Rs too closely, for we believe that drawing, games and singing are more beneficial and anyway the primary school will offer enough opportunity for the three Rs. The managing of the garden's fruits, leaves, shells, stones, etc., more readily captures the children's attention. Singing and dancing are daily occurrences. The idea behind these two activities is to help preserve and promote the vernacular and local culture, since in the ethnic communities of Indonesia, of which there are many, vernaculars and their attending cultures run the danger of disappearing, the more so now, with the national language rapidly gaining ground and popularity due to its being identified with progress. Artistic Bali offers a wide range of local instruments which can be used in kindergartens and are inexpensive compared to instruments that are imported. In our TK we have a few wooden xylophones on which the kids can beat to their hearts' content.

Hygiene and Clean Habits

Holistic or integrated health is a supreme consideration in our approach. Consequently health of body, mind, and soul should be promoted whenever and wherever possible. But the stimulation should take place in a natural way. For instance the singing of *mantras* before starting the morning and when leaving the premises as well as before partaking of lunch is our way of getting involved in religion and a reverent attitude. *Mantras* promoting love for parents, teachers and guests, or for instilling a feeling for God as being the essence of life as so well expressed in '*tvameva mata, cha pita tvameva*', or for promoting friendship, fill the Ashram atmosphere at set times of the day.

Keeping the little hands busy at something useful and meaningful fosters healthy minds, whereas bodily health is taken care of by exercise, walks, swimming and games. In order to stimulate them in a natural way it is especially important that the teacher teaches by example and by personal participation in these activities.

An important feature of the hygiene and health concern is the free school lunch we offer our children. It is through the school lunch that many things are passed on, such as: (a) nutritious food for our lunches is so planned that the parents can make the most of whatever the environment offers in the way of greens, fats, protein, etc.; in fact their habit, dictated by poverty, of eating a variety of green leaves from hedges nearby proves to be most wholesome. It has been found that, for instance, the consumption of three different kinds of green leaves as mixed vegetables more than matches two carotene intake we get from carrots, which are alien to villagers anyway besides requiring 'sophisticated' methods for their cultivation; (b) fermented tapioca cake is an excellent way of taking in vitamins B-complex, while such cake is very cheap and thus considered a 'poor people's snack'; (c) *tempe*, a fermented soya bean item, contains the best and most easily digested vegetable protein. At first it was rejected by the children, as it originated in Java, where it is a 'people's food item. Persistent persuasion has now made *tempe* acceptable and popular.

Neat eating habits should be established and brushing teeth afterwards is to habituate the children to a clean feeling in the mouth. From remarks among the parents when they bring their kids to school we can conclude that our free lunch programme does not stop at lunchtime in our premises, as every newly introduced food item is heatedly discussed. The pros and cons are debated, but it is certain that there are always a few who adopt our novelties.

Our dispensary, which serves the schoolchildren's health without excluding others from the village, has also helped in lessening the habit of smoking, for no smoker is served. Of late we have begun to offer acupuncture. The hardest fight was against the national habit of the Balinese of spitting. Luckily, this habit is dying out in our neighbour, the village school. A five-year campaign about plastic littering has yielded results, but not at the public level. That is, now people are careful about keeping their compounds plastic free, but this feeling of responsibility for public places is still very thinly spread.

The School's Interaction with Others

Interaction forms the basis of social intercourse of human society. It may be said that perhaps interaction beyond the family circle starts for the child in kindergarten. For the TK 'Gandhi' interaction has an added dimension. Naturally the children's interaction with Ashram members has its merits as well as a negative side. But an unusual opportunity is there due to our foreign guests' presence. Not a few of them took an interest in our TK kids and more often than not joined in by teaching simple songs and games from their own cultures.

Guests with small children loved to have their own offsprings join the kindergarten while they were holidaying with us. In the first year of the kindergarten children were afraid of our foreign guests, a few even immediately started crying at seeing a foreigner, but now we have to take care that they do not bother the guests, because of few of them like to spoil the children with all kinds of little presents, which can easily develop into some form of bribery. This aspect of interaction is conducive to developing the sense of 'one world' and the beginning of appreciation of cultures other than our own.

Instilling a sense of community is done through the children's participation in works of construction, repairs or building a new cottage. On such occasion the little kids can be seen joyfully carrying bricks or buckets on their heads and walking in single file to the site of construction. The weekly general cleaning-up of our premises always affords great hilarity as this gives them an occasion for running about everywhere, while not only collecting plastic rubbish but also firewood for the kitchen.

Interaction with nature at this tender age could be decisive for a child's later development. Our Ashram grounds and situation offer such interaction in abundance. Natural vegetation, stretches of green lawn, the beach and blue sea only a few metres away, the fresh pond bordering the Ashram grounds on one side, animals such as cows, calves, dogs, cats and a chicken or two, provide another living aspect of nature which may easily awaken a sense of wonder in the young child. And what is life worth without this sense of wonder?

Before school starts children take turns in tending the school garden, watering plants, pulling up dead parts; thus early they are trained to take an interest in whatever they do. Especially in gardening it is always an uplifting experience to watch a bean forcing up its way through the soil in a day or two and see it daily growing to full maturity, followed by the excitement of picking it and having it served as lunch afterwards.

The Place of Nature and Culture in Our School

Nature is divine for the Balinese. The primordial, deeply inbuilt respect and reverence for nature of the Balinese was a ready substratum for the *mahavakya* or great utterance in *Vedanta* introduced by the Hindus to Java and later Bali, in around the fourteenth century. This utterance *tat tvam asi* has even more strengthened our deeply-felt awe and reverence for nature, as is manifested on Plant Day, Animal Day, Tool Day, culminating in the feast of Sarasvati, the goddess of learning and the arts, and Earth Cleansing Day, which comes on the eve of our new year. These ceremonies, except for Earth Cleansing Day, come around every seven months, making Bali an island of daily festivals if individual birthdays and temple anniversaries are included. As a school we especially make much of Plant Day,

Animal Day, Tool Day and Ganapati Day. The most festive is Sarasvati Day, celebrated by all schools and institutions or organizations engaged in artistic and literary pursuits. On that day a priest is invited to lead the function of worship. Sweet and intimate is the celebration of Plant Day. On that occasion a good, fruit-bearing coconut tree in our garden is selected and decorated by the children. They come in their best dress to the school, each with a little offering of fruit and flowers. Sitting around the tree they sing devotional songs, glorifying and blessing the plants in the garden. After the prayer and *mantras* the kids and the decorated tree are sprinkled with holy water. This and the partaking of fruit and cakes in the offering is enjoyed by all.

Days before a ceremony takes place one can feel the increased activity in the air, for everyone is up and doing something. So it is in our school, as the children have to make the decoration and offerings themselves while only for intricate designs do adults come in to assist. With frequent celebrations, hardly a week goes by in our Ashram without this joyous and busy atmosphere.

Thus reverence and identification with plants, animals, in short with the whole of creation, are fostered in the young child. Gradually through devotional songs, dignified language and a respectful attitude towards creation, culture is instilled in the young souls. Our school celebrations culminate in the commemoration of Gandhi Jayanti on 2 October each year. Then the whole day our place rings with laughter, chatter and the joyful shouts of some two hundred or so kindergarten kids coming from several schools of the region to celebrate with us. Competitions are held in drawing, singing local children's songs and reciting *mantras* especially selected for young children, games and various skills such as cutting palm-leaf decorations, making little offerings, dancing and drawing. Viewing a whole year from this stance, indeed life for the Balinese is centred around celebrations. The whole of life in Bali is a celebration.

Daily Schedule of the School

The time children spend with us is from 7 in the morning till 11. During the four hours that they are under our care a general pattern as indicated by the time-table below is followed as much as possible:

07.00 - 07.30	arranging their tables and mats to sit on and sweeping the school garden
07.30 - 08.00	morning gathering with singing of *mantras* to start the day followed by local songs

08.00 - 08.30	outdoor physical activities such as games where bodily movements are emphasized.
08.30 - 09.00	swimming and bathing in the sea followed by rinsing in the fresh-water lake and going for a walk in the hills or watching the daily village activities.
09.00 - 09.30	drawing or practising manual skills such as cutting palm-leaf fronds, mat weaving, moulding sand in various shapes and sizes.
09.30 - 10.00	serving lunch and washing up food bowls.
10.00 - 10.30	tooth brushing, as we cannot rely on parents to have their offsprings do this daily.
10.30 - 11.00	story-telling by teacher and children alternately.

However, this time-table is quite flexible, for whenever something special takes place that event or occasion is given priority. The children are most of the day outdoors and only drawing is done inside the classroom. Although in the curriculum prescribed by the government the time devoted to the three Rs is prominent, more importance is given by us to free drawing with chalk on blackboards cut to size or on the sand of our beach. This has been adopted to save on materials for which we would have to spend more money. Learning to discern letters from the alphabet is done by way of drawing enlarged letters, the arrangement of which should teach them to spell their own names and those of the most conspicuous objects in their environment. In this context the weekly walks in the hills and the neighbourhood serve to stimulate their imagination and power of observation, which afterwards finds expression in the drawing class.

Interaction with village institutions happens largely through the village chief, and it goes without saying that our channel to the island's government is the district supervisor. The parent-teacher association we formed at the beginning became inactive after a year due to the failure of parents to attend meetings. To make up for this, we have our teachers make home visits at least once a week in order to familiarize themselves with the home conditions of the children.

Most parents have such a struggling existence just to meet the family's daily needs that they are unable to get involved in school concerns. It is well-known that poor parents delegate all the responsibility for their children to the teachers and find a happy excuse for this by saying that they are ignorant and backward people who cannot be expected to think beyond scraping a living.

After 15 years of running TK 'Gandhi' the following observations can be made:

1. Running a kindergarten in our area also means endeavouring to change the parents' outlook on life.

2. How difficult it is and what a long time and patience it takes to break bad habits and establish new ones in their place.

3. How good and sensible eating habits, which includes some understanding of nutrition, ultimately trickle through to parents, be it at a snail's pace.

4. If at first it took some cajoling and persuasion to draw children to our school, now there is no need to do that. The real problem is how to cope with so many children with only two teachers available.

5. How hard it is to stem the tide of consumerism and to correct the idea of 'progress'. Especially the newly developing countries are dazzled by material progress as displayed by the West. Whereas our orientation is human progress, which is anchored in the spirit unanimously pronounced in all scriptures, such as the Bible's admonition: 'What profiteth it man to gain the world and thereby lose his soul?'

6. One particular phenomenon with poor people dies hard: as soon as they have a bit of money it is spent on status rather than on necessities. This has been driven home to us very clearly in the TK 'Gandhi'.

   We have tried very hard to stick to simplicity and truth. Thus we have never encouraged the wearing of shoes and socks, nor of a uniform. Instead a pair of thongs will do for the protection of the children's feet, while on our premises the children are encouraged to wear as little clothing as possible. When a different group of people came to settle down in our area, that is when it opened itself to tourism, the children of this group of clerks and hotel employees entered our school. That happened in 1986 and shoes, stockings and town clothing made their entry into our TK. Uniforms copied from other kindergartens appeared. Not long after, even the poor fishermen and farmer parents insisted on their children's wearing the same outfit, even though they had to borrow to meet the expense. We have not yet decided whether we should bar the wearing of uniforms in our school.

7. Another problem is the increase of 'snacking' in the Ashram grounds, and what annoys us most is the amount of fast food snacks the children bring with them

from home. Naturally they are mainly the children connected with hotels, restaurants or shops. We are thinking of being more strict in this regard with the new school year by prohibiting the bringing of snacks into our premises.

8. A more optimistic note, luckily, can be sounded. When during the first three years we started school with giving the children baths, paring their nails and showing them how to use the toilets, etc., this can now be dropped. Spitting and littering have almost disappeared, while whatever food is served at lunch is neatly finished and food bowls properly washed.

9. We started the school with 27 children, painfully gathered over three months. For three years this number remained more or less constant, then dropped to ten kids in 1983-84. Most likely this was a reflection of the fact that by that time family planning had already taken effect and small children were indeed scarce.

In 1986 the number picked up, for the children of tourism-employed parents flocked to TK 'Gandhi'. This changed the aspect of our school somewhat and the town image gained the upper hand. If the local children became somewhat naughtier after a year of school feeding, now their general behaviour became distinctly rougher, their language at times rude, while some among them have become downright cheeky. This year, with 40 kids to cope with, our two teachers have their hands full and two other Ashram members have had to be put on duty to maintain order and the general functioning of the school.

From the above it is clear that the reader should not be misled by the term 'kindergarten' evoking the picture of a well-furnished and well-equipped classroom with teaching aids and all the paraphernalia common in the West. All these we cannot afford. From that point of view perhaps our venture should be termed a 'child-minding' service. However, the spirit that moves our undertaking should be seen as part of an endeavour to free ourselves of foreign domination. It is the spirit to hold on to our true self and identity, to preserve our mother tongue, songs, dance, crafts: in short, all that is our own and has in course of time given birth to our particular identity. Yet we must do this without being deaf or closed to what comes from outside. I must quote here Gandhiji's incisive remark: "I do not want my house to be walled in. Let the winds from all corners blow freely in my house, but I refuse to be blown off my feet." To enable us to stand firm on our own feet a start should be made in kindergartens when the mind is still tender and malleable and emotion still relatively free from self-interest and calculation. Pondering over this I realize what tremendous patience and devotion are required from kindergarten teachers. A kindergarten is, in my opinion, a place where *ahimsa* can be practised at its best. Again, as Gandhiji observed: 'He who has not in him the quality of infinite patience cannot observe non-violence'.

Prayer

Learning, a joy

Playing with the mud

Making a musical instrument

Creativity
hance happiness

A world of innocence

The Smile of today . . . .
Hope of tomorrow

The Childhood innocence

Playing
together

The self-activity of the child

Teacher stimulating the child

Painting is writing

Celebrating Gandhi Jayanti

For Gandhiji *ahimsa* and active, pure love were interchangeable terms. We are aware of the very tiny speck of progress all our fifteen years of existence have shown. Even to achieve this all our Ashram members — in all, some fifteen girls and women — and our many guests and student volunteers have been involved and have interacted with our TK kids. In view of the present global problems it is a most pressing need to provide sound and inexpensive education, at as little cost as possible, in order to remain within the reach of the masses. And another remark of Gandhiji's will help us to search unceasingly for the application of *swadeshi* in establishing our own identity. He said lucidly and pithily: 'What the average person cannot have I should refuse to have'.

Plans for the Future

We have nursed the hope some day to have our own primary school, a follow-up to our efforts in the TK to lay the base for clean, healthy, and creative living. It is said that whatever good habits we managed to establish among our children during a year most likely will suffer abrasion after years of attending other schools later. One year is too short a time to achieve anything solid. But circumstances and conditions make this hope a very distant dream. Apart from financial limitations, to start another primary school is also superfluous when the existing two now find it hard to fill their first grade classes. But we should and must continue improving and upgrading our present undertaking. Our present experience could be put to good use in what we plan to start next year. Our 1996 project is named 'Holistic health to our countryside'. In this project the upgrading of kindergartens will also be covered.

Having run our public reading room at Canti Dasa Ashram for three years we have found that it does not satisfactorily meet our expectations, i.e., to promote and stimulate healthy reading habits among village youth and avail of the opportunity to offer a reading place to the adjoining village school of Candi Dasa. Therefore we have decided to try a somewhat different approach with our new project.

The main idea behind this project is to promote holistic health in the countryside, particularly addressing ourselves to the youth and adolescents. Holistic health will concentrate on offering classes, free of charge, of the following nature:

(a) Teaching *Hatha Yoga* for the promotion of bodily health in a manner, which requires no expenses.

(b) *Mantra* chanting, promoting mental peace and a desire to know those *mantras* which strengthen will-power, solidarity, the shedding of fear, and in general promote a non-violent attitude and turn of mind.

(c) Promoting familiarity with and desire to live up to Gandhian thought and life-style.

(d) Nutrition in conjunction with *rshi-kriti* or non-violent farming and gardening.

(e) Revival of local nursery rhymes and children's songs.

A sizeable van manned by Ashram members to teach the above classes, also carrying edifying literature for young people to widen their horizons without upsetting them, will move from village to village according to a schedule. For three consecutive months two villages will be alternately dealt with, in the hope that in the course of five years 40 villages will have been exposed to our classes. The morning hours should be devoted by the Ashram team to gaining information on eating habits, hygiene in general, existing local skills, crafts and customs, since our classes will be held in the afternoons. This information should serve us well in carrying out our project. Socializing with established and influential village members should be given prime importance.

Every weekend the team will discuss the information gained and the week's experience in our Ashram, their base.

We believe that by faithfully carrying out this project for, say, five years, we will have reached enough young villagers to spread general knowledge regarding 'holistic health', which covers sustainable nutrition based on chemical-free food, nature cure supported by yoga, chanting and acupuncture and taking up handicrafts as the main cure for 'woolly' thinking and idleness.

All these activities should lead to greater self-esteem and self-reliance, because they are rooted in the sound faith of religion.

The first and indispensable requirements for our project are therefore a sturdy van and devoted, skilful teachers. To obtain these is now the focus of our effort and energy.

M.K. Gandhi (1869-1948) and Ki Hadjar Dewantoro (1889-1959)

If in India Gandhi was unanimously hailed as the Father of the Nation, in Indonesia Ki Hadjar Dewantoro is looked upon as the father of national education and culture. Gandhiji's life could be said to have been devoted to the welding of all the ethnic communities in India into one nation that could be likened to a huge tapestry of brilliant colours, a reflection of the many customs and cultures of its components.

Indigenous education in Indonesia were the *pesantren* for the Moslems and the *ashrams* for the Hindus. Both were largely occupied with religious instruction. With the advent of the

Dutch, who introduced their own system of education, the Moslems retreated even more into their *pesantren*, making of them a stronghold for the education of their youngsters. But in Java, where the indigenous orientation called 'kebatinan' was deeply rooted, *pesantren* became strongly coloured by this ancient belief, which is a kind of mysticism. Ki Hadjar Dewantoro, a prince from a court in central Java, underwent the same agony as Gandhiji when noticing the suffering of his people under the Dutch heel. After several years of political life and pamphleteering he arrived at the same wisdom as Mahatma Gandhi did: that the road to the real freedom of his country could only start with education. Without completely dropping his vocation as journalist and political activist, he translated his ideas on education into action and started his now well-known institution of education named Taman Siswa. It literally means 'a garden for children'; but the clear connotation is 'a place to learn how to live'.

Twenty years separated him from Gandhiji, but the same ideas, by and large, kept both men struggling and fighting foreign rule throughout their lives in a down-to-earth manner, through opening schools, starting cooperatives, encouraging all kinds of productive activities. Both suffered imprisonment. Ki Hadjar Dewantoro even was exiled to live in cold Holland far away from the 'field' which he wanted to cultivate. Living there in near penury with his wife and two small children, his fighting spirit or his political activities did not abate.

After six years of internment the Dutch, forced through circumstances, had to release him. With the financial help of friends he was able to return to his beloved Indonesia, where undauntedly he pursued his struggle and opposition to the Dutch colonial government. The Japanese occupation came, and according to the Japanese, now that the Dutch were gone there was no need any more to have 'national schools' side by side with government schools. However, in an unobtrusive way Taman Siswa education continued to be pursued, for Ki Hadjar Dewantoro, with his clear vision, saw that its spirit must be maintained and fostered in the teeth of the existing schools, which were fed on pseudo-nationalism while in fact preparing us for another yoke, this time the yellow yoke.

This is not the place to elaborate on Ki Hadjar Dewantoro's philosophy, which forms the basis of Taman Siswa education. Nevertheless a few outstanding ideas found in his philosophy should be mentioned to see how parallel Gandhiji's and Dewantoro's ideas run. This is not to be wondered at, because both selfless and patriotic men were passionate lovers of freedom, not only for themselves but for all the people they belonged to. Thus both were the embodiment of their peoples' aspirations and suffering. If the motives are the same, the outcome and method of the struggle must be the same. The more so, because both had an unshakeable faith in the power of God.

Freedom-fighter and author of national education and culture, Ki Hadjar Dewantoro's vision affected not only the world of education but also social life and society itself. The following points show clearly where both the great men, Ki Hadjar Dewantoro and Gandhiji, met:

(a) Cultural nationalism, a favourite term with Ki Hadjar Dewantoro, has been accorded a high place, which shows his ability to look to the future. For the many and varied ethnic communities of Indonesia must arrive at a kind of national culture to weld the country into one solid cultural region.

(b) He revived the phrase 'Tut Wuri Handayani', an ancient Javanese aphorism expressing an educational principle in teaching and bringing up children. It gives full freedom to the child's initiatives, while parents follow from behind ready to correct it when things tend to go wrong or cause harm. In a wider sense it applies also to the relationship between government and governed, or state and people.

(c) The Taman Siswa philosophy believes in the right of self-determination.

(d) Taman Siswa adheres also to democracy, but one with firm moral leadership which keeps it from deteriorating into unbridled freedom and anarchy.

(e) It upholds self-reliance which, consequently, imposes voluntary simplicity, planning and healthy discipline on Taman Siswa people.

(f) The extended family approach — so natural among Asians — should be observed and applied in economic, social and political spheres as well.

(g) Principles of concentricity, convergence and continuity should be honoured. These three Cs require some explanation, for Ki Hadjar Dewantoro attached a special meaning to them. Concentricity in the Dewantoro sense fully agrees with Gandhiji's idea of what he called 'concentric circles' as seen in the ocean. Thus a person, apart from being his personal self, is also part of the family and of mankind. The term 'convergence' as used by Ki Hadjar signifies the relationship that should exist between small, big and biggest communities. Finally, by continuity he referred to the unbroken nature of culture as it developed from the hoary past.

(h) Nature and her laws reign supreme in the Dewantoro world-view as they did for Gandhiji. We therefore go to nature and observe her, learn from her how to live in harmony with her and with ourselves.

These points require unceasing observance if we want to see their result. That Dewantoro's view was prophetic is supported by the fact that the Taman Siswa schools have multiplied, not only in Java, but also in Sumatra and other islands in Indonesia.

The democratic character of Ki Hadjar Dewantoro is patent. Born of an ancient noble family, without hesitation he dropped his title of nobility — no mean thing in those days (1913) when nobility was much sought after by Indonesians — and changed his name from Raden Mas Soewardi Soerjodiningrat, in England the equivalent of Lord Such-and-Such, into plain Ki Hadjar Dewantoro in order to prove his belief in a classless society. Whereupon the whole Taman Siswa world followed suit, adopting the term Ki for men, Nyi for married women, and Ni for unmarried girls.

His view of nationalism should be made clear, now that here and there nationalism tends to narrow down to fanaticism and exclusivism. Nationalism to Ki Hadjar Dewantoro should not violate or go against the humanitarian universal values we find embedded in all world religions. Also, his rejection of idolizing human beings while they are still alive or putting them on pedestals is worth noting. He shared this belief with M.K. Gandhi, who was averse to any kind of human idols.

Ki Hadjar Dewantoro's spirit of dedication to the child is preserved in the vow every Taman Siswa Teacher has to take:

> Free from any form of attachment, in the purity of my heart, I will approach the child entrusted to me. Without demanding any right or privilege instead, in all fullness I will dedicate myself and serve it to the best of my capacity.

Is it to be wondered then that his birthday, 2 May, has been declared by law on 16 December 1959 as a national holiday under the name of National Education Day?

2

Poverty and Education
The Samanwaya Ashram

Dwarko Sundrani

ONE morning in September 1968 I went to village Bharatpur, 6 km east of Bodh Gaya, to bring a Musahar child to our school. Bharatpur had been established by us on a plot of 100 ha of land on the other side of the river Niranjana. The land had been donated by the Bodh Gaya Math. The villagers suggested Dharmanath, an eight-year-old boy, for schooling in the Samanwaya Ashram. I was returning with Dharmanath, an orphan from the poorest of the poor Musahar community, with an Englishman, Alan Lether. The Niranjana, which had witnessed the historic moment of Siddhartha's becoming Buddha, was witnessing yet another event. We were crossing the river and when we were in midstream, about fifteen persons armed with sticks (*lathis*) and led by a landlord, Bhawan Singh, came to beat us because we were taking Dharmanath for schooling.

Dharmanath had committed the sin of being born a Musahar and the landlord had paid Rs. 6 to perform the last rites of his mother. Dharmanath could not pay back that money and thereby became a bonded labourer for life. Bhawan Singh hurled abuses and wanted to beat me because if Dharmanath began studying he would lose a labourer. My whole philosophy of non-violence and peace was at stake. I gave many arguments against the system of slavery and labour but the more I argued, the more angry Bhawan Singh and his goons became. Mr Lether had taken a photograph when they wanted to beat me and had run away with his camera lest it should be snatched from him. Some of the men feared grave consequences because of the evidence of beating taken by camera. The landlord and his men went back. Dharmanath was so frightened that he ran for his life. I returned to the Ashram, without Dharmanath and with a heavy heart.

Our Ashram had been inaugurated on 18 April 1954 by Shri Rajendra Prasad, the then

President of India, in the presence of distinguished persons like Pt. Jawaharlal Nehru, Shri Jayaprakash Narayan, Acharya Kripalani, Professor S. Radhakrishnan, Shri Gulzarilal Nanda and others. The Ashram was founded under the loving care and guidance of Acharya Vinoba Bhave, with Shri Kaka Kalelkar as the first president and myself as the managing trustee. It was established to bring about social change and awareness in Bodh Gaya, one of the most backward regions of our country, mostly inhabited by Scheduled Castes and Scheduled Tribes. Poverty and deprivation are the destiny of the people of this region.

One day we were passing through village Kadal under Barachatti block with a Frenchman driving our jeep. All of a sudden a Musahar woman in a tattered cloth which could hardly cover the private parts of her body dashed over to our jeep and started narrating her harrowing story. She stated that she had not eaten anything for six days. Her husband had died of snake-bite six months back. She had three daughters and it was becoming very hard to feed them, what to talk of education. I gave her Rs. 5 and enquired how long she would feed herself and her three daughters with the money. She said that it would be sufficient for five days. I was so overpowered by Lord Buddha's compassion that I decided to start a school exclusively for scheduled caste and scheduled tribe boys and girls. I begged the woman to give one of her daughters for making my experiment with poverty and education, to which she readily agreed. The daughter was named Sita and thus Samanwaya Vidyapith was established on 15 June 1968. We decided to make a new experiment with the Gandhian concept of education. We called upon the Mahant of Khajwati for a donation of land for this educational experiment. He readily agreed and donated 31 acres at Bagha near Barachatti block, 35 km south-east of Bodh Gaya.

One fine morning when I saw Dharmanath entering our Vidyapith, I could not believe my eyes. He was a changed boy. In the past three months he had undergone a great metamorphosis. I asked him whether the landlord would again start beating me when he discovered that he had come for schooling. His answer was so blunt that I could not believe it. He said that if the landlord ever tried to snatch him from me, he and his goons would be murdered by Naxalites. Only then did I know that Dharmanath had been sent by left extremists who had established a firm hold over his village.

Poverty breeds violence and hatred. Education breeds goodwill, peace and harmony. Without proper education and social awareness the establishment of an egalitarian society is a distant dream. Bodh Gaya presents a paradox — ignorance in the land of Enlightenment. In spite of the fact that government had declared education compulsory and free forty-five years back, not even 2 per cent of Musahar boys attend school even today. The percentage of Musahar girls going to school would not be even 0.01. The government's declaration of education for all up to AD 2000 is nothing but a mockery of education and democracy.

Musahars are the most backward community among scheduled castes and Bhoktas are the most neglected scheduled tribe in Bihar. Musahars are landless and born bonded labourers. They have been exploited for such a long time that they consider starvation and deprivation as their only destiny. In fact they are living in sub-human conditions.

The culture of poverty which they breathe has left deep scars on their personality and behaviour. They show no initiative in any work. They are totally submissive and agree to any suggestion given by any authority without applying any reason of their own. To think of values, social, moral or spiritual, for them would be unrealistic. Musahar society suffers from deep inertia and depression. The population of Musahars is 30 million in Bihar. The Bhoktas are fewer, with 100,000 people. They live in the forest by hunting and eat fruits, roots and leaves because they do not have food even 48 years after Independence.

The Samanwaya Ashram started work to uplift these communities in 1954. The land given by the gift-movement (*bhoodan*) was distributed among them. 5,000 acres of land were distributed amongst 3,000 families in this area and 13 new villages were established. The Samanwaya Ashram worked very hard to develop these villages, but due to the poverty and ignorance of these communities, progress was not encouraging. So in 1963 I went to make an experiment in one village, Manfar, among these two communities in education and development. I collected 29 young boys, who were herding goats and pigs, for education. These boys used to come to me after dinner in the night and study for one hour. After prayers they used to go to sleep, and early the next morning after prayers they would study for one hour and then go home to work. This experiment ran for one year but I was not satisfied. The children were studying for two hours a day but this did not bring durable change in their attitudes. I came to the conclusion that if we want to bring about real uplift, we need residential schools.

I approached the Bihar State Government to establish a residential school in Manfar village. Government approved and started the school in 1964, but there were bureaucratic problems. After I brought these to the notice of higher authorities, the government took the decision to appoint me the secretary of this school, whereas in the whole of Bihar there were government officers as secretaries. Unfortunately I could not succeed in running this experiment, so I resigned.

During the Bihar famine of 1967-68 the Samanwaya Ashram did relief work in the area. I came in close contact with the weakest sections of society, especially Musahars and Bhoktas. Although these people were starving, they would not work. Ignorance and illiteracy were the main causes of this inertia. I found that the greatest sufferers were young children, so I decided to work for the education of these neglected children in a residential school.

Education has been of great concern for every President and Prime Minister of India.

Recently the government published a document called *Challenges of Education* in which it has been accepted that the present system has proved to be a great failure. In spite of many commissions, nothing substantial has been done. I feel that it would not serve any purpose by condemning the system. It is necessary to find the solution, so I started to search for an alternative educational system. The Samanwaya Vidyapith is the result of my experiment with poverty and education.

In 1971 Shri Jayaprakash Narayan, the great leader, visited the Samanwaya Vidyapith and remarked,

> Samanwaya Vidyapith is a dream come true. Dreams hardly ever come true, but it is the genius of one man and his capacity for experimentation and his innovation and his undiminishing ability to learn from other experiences that have combined to create this splendid institution — that man is Dwarko Sundrani.

It is said that nations are not built by parliaments or assemblies. Nations are built in family and in school. As Lenin also said, "If I know the songs on the lips of your children, I can tell you the future of your nation". People's education and child education are interdependent and they must work simultaneously. I am in search of education of life, education for life and education through life. An education can be summarised in three words: *yoga*, *udyoga* and *sahayoga*: total development of life, i.e., physical, emotional, intellectual and spiritual; correlation of education and manual work leading to self-reliance in the basic necessities of life; and harmonious living with nature and society. The aim of education is not solely to induce literacy and numeracy. This is only a medium. The aim of education is to develop the whole personality of the child.

The personality of the child is developed in the family and in the school, hence it is essential to develop the family too. People's education and children's education must go hand-in-hand, so education must be correlated with development work. Our first Prime Minister, Shri Jawaharlal Nehru, had said, "A poor country like India needs a school which would be a centre of development also". The Samanwaya Ashram has taken the children of Musahar and Bhokta families for this new experiment in poverty and education. The families are helped in the villages through development work and the children are educated in the Samanwaya Vidyapith. After education the children are established on the land that is given to their families.

In India there is little education in the rural areas, as these have suffered neglect for centuries. If the village child is educated, he tries to find a job outside the rural areas and is uprooted from the family and the village: there is a brain drain from rural to urban areas. But India is a country where 80 per cent of the people live in villages. How can these people get jobs in the urban areas? Education should help them to stay in the villages.

The Samanwaya Vidyapith is correlating education with agriculture, dairy farming and mechanics. In mechanics, diesel engine repairs, electrical motor winding and jeep-driving and such other training is imparted. The masonry needed for low-cost housing is also taught. The education is also combined with singing, dancing, painting, puppet shows and prayer, etc.

Generally the purpose of going to the school is to get a diploma and eventually a degree and a job. In the Samanwaya Vidyapith no certificate is offered. There is no government grant, no government certificate, no government books, no government examinations. The ability to earn and learn is developed. There are about 100 boys and girls in this experiment. The age is from 5 onwards, and the Vidyapith takes the responsibility after education to marry these children and settle them in the villages. After this had run for 20 years the Director of the National Council for Educational Research and Training (Non-formal Education) in Delhi came to see the experiment. He was very impressed and pressed me to work with the government. As I do not have good experience of government working I did not agree, but after pressure from him I agreed to do extensive work in villages.

The Samanwaya Vidyapith is laboratory work and village education is extension work. There are 86 villages where 150 centres are being run. There are 4,577 pupils getting education. In seven villages every child is taken in the school. In these villages education is also correlated to village problems. Due to deficiency of vitamins in the food there is night blindness in these villages, so we give each child ten green vegetable seeds and three papaya trees to grow. The villagers are also selling vegetables and fruits and earning money. Cholera is endemic in this area, so the children and the teachers go and disinfect the drinking water. They make compost manure from the waste in the villages. Lessons in language, arithmetic and science are based on growing vegetables and fruits, making compost, and health matters. The government is giving Rs. 800,000 per year for this project and these centres produce Rs. 700,000 from vegetables, fruits and manure. We are proud to express that we receive full cooperation from government agencies in this work. It has made good progress in the integration of religious sentiments. Seven Muslim teachers are teaching the *Ramayana* to Muslim girls on their own initiative, and the Hindu teachers have started teaching the Holy *Koran* to Hindu children. This area is one infested by terrorists. Our educational system has had a good influence on the youth in the area, with the result that 25 young terrorists have been converted and have joined the programme.

When I am asked about the results and achievements of this experiment, I reply that this experiment did not start from zero. This experiment was started from minus 40 and has not reached zero yet. So one cannot see the results, but if one knows the background from which these children have come, one can anticipate the likely outcome. There cannot be reports of such experiments; one can only experience them. There is definitely a change in the attitude of the pupils when they go home. We have prepared teaching material suited to the home environment of the children. We have also prepared training materials for the teachers.

As regards the financial aspects of this experiment, we are glad to say that we have never asked for any financial help from any individual or agency, and for 100 resident children and 4,577 day scholars, we have never felt any financial difficulty. Money is coming and the work is going on smoothly. There are 100 sponsors from different parts of the world. Each sponsor sends US $10 per month per child. And we also produce food, etc. on our farm.

People ask about the problems that this experiment has faced. It has been a long journey of 25 years. We have faced many problems but they did not hurt us or dishearten us. There are no monetary problems, no administrative problems. There is only one main problem in this experiment. There is a dearth of good, efficient and committed teachers. Such teachers are not available.

Finally, the name of the institution is Samanwaya Vidyapith. 'Samanwaya' means harmony. The objective of this institution is to bring harmony. At present we are passing through a period of transition which is unprecedented in the annals of human history. Man is in the crucible. He is facing a curse, a threat to his very existence on the one hand, and the blessing of an opportunity for a rebirth on the other. Harmony is the solution. There is necessity to give education in harmony. Harmony can be established only through mutual understanding. Mutual understanding can be created through service to one other. The Samanwaya Vidyapith is working on these lines. The poor children are being educated without any caste, colour, creed or religious considerations. They work together, they serve together, they live together. The haves are sharing with the have-nots. Hindus, Christians, Muslims, Buddhists and Parsis are involved in this experiment. Hundreds of families from different corners of the world are coming to the Vidyapith and extending cooperation. This is the faith of serving and helping one another, which will harmonise our existence. This was the dream of one-world and one humanity of Gandhiji. The Samanwaya Vidyapith stands for the education of the masses and not of the classes. It is through education that we can establish a classless and casteless society, which is the need of the hour. So long as there exists a Dharmanath or a Sita, any talk of new education, socialism, democracy and egalitarian society are hollow slogans to befool the Indian masses. The situation is explosive. Any further neglect of such children will destroy our social fabric and engulf us in chaos, disintegration and decadence.

3

Rural Context of Primary Education

Searching for the Roots

Shakuntala Bapat & Suman Karandikar

THE present system of education in India, from the preschool stage to higher education, has been imported from the West in bits and pieces over the last 200 years. The overall cultural contexts of Indian society and the cultural specialities of its varied segments have been ignored by this system, with the result that it has never been fully accepted by the people. It hardly needs to be pointed out that the ecological inheritance, ethos and cultural commitment of Western societies have been quite different from those of oriental societies. The climate, natural environment, types of settlements, their historical evolution and the resultant goals and occupations, and the life-views of these societies have always been poles apart. This is the main cause of the continuing discord between education and society in India.

That educational systems are subsystems inherent in any given social system and cannot be imposed from outside without damage to the social fabric was realised by Mahatma Gandhi long ago. In his speech delivered on 20 October, 1931 at the Royal Institute of International Affairs, London, he lamented that the British administrators of education had failed to notice the special characteristics of Indian culture which had a tradition thousands of years old of education and instead of taking hold of things as they were, they had begun to root them out. They scratched the soil and began to look at the root, and left the root like that, and the beautiful tree perished. The unrest which filled Gandhiji's heart at the sight of the mindless destruction of the vast network of literacy and knowledge-gathering enterprise embedded in Indian culture was directed not so much against individual officials but against the colonising mindset which always attempts to impose its own culture on a vanquished society. But this view of the colonial educational system was neither understood nor shared by many an educated Indian of those days. Gradually, however, this understanding did dawn on Indian educationists, but by then the damage had already been done.

The imported system had created two new classes in Indian society: the elite and the masses. The hybridised elite and the conquered masses became two cultural groups in a continuous state of conflict in the socio-economic and political fields. Efforts to cope with this conflict have seriously told upon the people's intellectual health, creative urges, self-respect and self-confidence. This damage to the Indian spirit has to be understood if a new educational era is to begin for the Indian people and give them the strength to meet material and spiritual challenges. The remedy was prescribed by Gandhiji: searching for cultural roots which have a refreshing diversity along with a common bond forged over thousands of years, and appreciating the beneficial aspects of other cultures with an open mind.

Indigenous Education

The indigenous system was studied by some British officials and scholars in the early nineteenth century. Even though their enquiry was restricted to British occupied territory, their reports serve to give quite a clear picture of the state of indigenous education even after the British had imposed their rule on most parts of the Indian subcontinent. There was an enquiry into indigenous education in Madras in 1822. An enquiry was conducted in Bombay Presidency in 1823 through the collectors of districts. In Bengal, Lord William Bentinck ordered an enquiry which was carried out by William Adam, a missionary who took a keen interest in Indian education. The Madras and Bombay enquiries were rather rough jobs. Adam, who studied only five districts out of nineteen in Bengal and Bihar, had an inadequate sample; but his work was thorough and his reports surprisingly unbiased.

The Madras enquiry found that there was one native school per 1000 population but there were hardly any female pupils enrolled. The pupils were generally between the ages of five and ten but many boys continued up to twelve or fourteen. There was a large practice of domestic instruction, and the number taught at home was five times greater than that taught in schools. Children were taught at home by relatives or private teachers. The report of this enquiry said, "the state of education here exhibited, low as it is compared with that of our own country, is higher than it was in most European countries at no very distant period. It has no doubt been better in earlier times" (*Selections from the Records of the Government of Madras*, No.II, Appendix E). The remark that the system had been better in earlier times shows the soundness of Gandhiji's judgement that more literacy and popular education prevailed in India before the British conquest.

The report of the Bellary Collector, who participated in the Madras enquiry, is detailed and gives a graphic picture of the indigenous elementary schools of those days. He states that children, mainly boys, were inducted into schooling at age five. The parents of the prospective pupils invited home the master and the boys already studying in his school. They sat in a circle around the image of Ganesha, the god of learning. The child to be initiated was placed exactly opposite the Ganesha image. The master sat by his side. After offering *puja* to Ganesha, the master caused the child to repeat a prayer to the deity, asking for wisdom. He then guided the child to write with his finger, in rice, the mystic name of the deity (*shree ganeshaya namaha*). The parents then gave a present to the master. The child began to attend school the next day. The initiation ceremony and the gift from the parents prepared the child's mind for scholastic work. Whereas legal compulsion for schooling existed in the West, in India it was popular tradition that prepared the parents and the child to accept schooling in India.

Most children continued at school for five years, although the parents withdrew some earlier due to poverty or other circumstances. But some continued up to 14 or 15 years. School started at 6 o'clock in the morning. While the first pupil to arrive was honoured, late-comers were punished. Idleness was not allowed. The pupils were divided into classes. The youngest ones were placed in the care of a monitor while the master himself guided the older or slower pupils.

Instruction began with the child writing the letters of the alphabet in sand. When his fingers were well-trained, he began writing on a wooden slate smeared with rice-paste and pulverized charcoal. Another variety of slate was made from cloth stiffened with rice-water and covered with charcoal and several gums. A pencil made of white clay was used to write on these slates. The writing could be wiped off with a wet cloth. These were inexpensive materials made locally by the pupil's parents or neighbours or older playmates.

After learning the letters, the pupil proceeded to learn conjunct consonants, vowel signs, the names of birds, trees, etc., and then began arithmetic which, starting from the counting of numbers, went up to fractions, measures of capacity, area, weight, and so on. The test of reading and writing consisted of deciphering various kinds of handwriting in a public performance. Letters and documents which the master collected were read out before elders. Writing letters and drawing up documents, committing poetry to memory with attention to clear pronunciation, and readiness to correctly read any kind of composition, were required achievements. Appreciating this teaching-learning system, the collector's report says,

> The economy with which the children are taught to write in the native schools, and the system by which the most advanced scholars are caused

> to teach the less advanced, and at the same time to confirm their own knowledge, is certainly admirable, and well deserves the imitation it has received in England.

In regard to the situation of indigenous education in the Bombay Presidency, the statement made by G.L. Prendergast, a member of the Bombay Governor's Council, in his minute to the Council, is extremely important for understanding the state of indigenous education in that vast region. Published in the Bombay government educational records, the minute points out,

> I need hardly mention what every member of the Board knows as well as I do, that there is hardly a village, great or small, throughout our territories, in which there is not at least one school, and in larger villages more, many in every town and in larger cities in every division; where young natives are taught reading, writing and arithmetic, upon a system so economical, from a handful or two of grain, to perhaps a rupee per month to the schoolmaster; according to the ability of the parents, and at the same time so simple and effectual that there is hardly a cultivator or petty dealer who is not competent to keep his own accounts with a degree of accuracy, in my opinion, beyond what we meet with amongst the lower orders in our own country; while the more splendid dealer, and bankers keep their books with a degree of ease, conciseness and clearness, I rather think fully equal to those of any British Merchant.

Also, the Fifth and Sixth Annual Reports of the Bombay Education Society (1819 and 1820, pp. 11 & 21, respectively) point out

> There are probably as great a proportion of persons in India who can read, write and keep simple accounts as are to be found in European countries . . . . . Schools are frequent among the natives and abound everywhere.

As to the situation in Bengal, Adam's first report stated:

> Indigenous Elementary Schools: By this description are meant those schools in which instruction in the elements of knowledge is communicated, and which have been originated and are supported by the Natives themselves, in contradistinction from those that are supported by religious or philanthropic societies. The number of such schools in Bengal is supposed to be very great. A distinguished member of the General Committee of Public Instruction in a minute on the subject expressed the opinion, that

> if one rupee per mensem were expended on each existing village school in the Lower Provinces, the amount would probably fall little short of 12 lakhs of rupees per annum. This supposes that there are 100,000 such schools in Bengal and Bihar, and assuming the population of those two provinces to be 40,000,000, there would be a village school for every 400 persons.

That India had widespread arrangements for schooling — not in the Western sense, but in the form of family-based instruction or learning centres conducted by a local instructor supported by villagers — is proved by evidence available from British records. As pointed out by M.R. Paranjpe,

> At the beginning of the nineteenth century, there existed a fairly widespread organization for primary education in most parts of India. In Madras Presidency, Sir Thomas Munro found 'a primary school in every village' (Mill, *History of British India*, Vol. I, p.562, 4th edition). In Bengal, Ward discovered that 'almost all villages possessed schools for teaching, reading, writing and elementary arithmetic' (Ward, *View of the Hindoos*, Vol. I, p. 160). In Malva, which was for more than half a century suffering from continuous anarchy, Malcolm noticed that 'every village with about a hundred houses had an elementary school at the time of its coming under the British suzerainty' (Malcolm, *Memoirs of Central India and Malva*, Vol. II. p. 150).

Adam found that there were Bengali, Persian and Arabic schools in the Thana of Nattore. The overall pay of elementary school teachers was Rs. 5 to Rs. 8 per month. However, education of girls was neglected, and among 30,915 scholars Adam found only 214 girls in 6 special girls' schools in Murshidabad, Birbhum and Burdwan districts.

These indigenous learning centres thrived well because they met locally perceived educational needs. Their timings, style of functioning, achievement system, the accountability of the teachers to the community, vacations and celebrations, and general flexibility, all were rooted in community culture. These important indigenous factors were ignored by the imported system. The government dominated formal system at first diminished and later completely wiped out the role of the community in elementary education. Consequently, the progress of elementary education in India suffered a serious setback, the problem of illiteracy arose, and an educational rift between urban and rural areas was created. The action research projects of the Indian Institute of Education, which have sought to spread primary education to out-of-school rural children, therefore, take into account the importance of linkages between community culture and education as their main foundation.

Culture-conscious Primary Education

Glimpses of indigenous education as seen above, and many recent research studies carried out by Indian and foreign scholars, have clearly shown that universalisation of primary education in India's rural and tribal areas, which contain nearly 75 per cent of India's population, has been hampered by poverty but to a larger extent by the inappropriate organisation of the education system, which has persisted in ignoring the cultural contexts of the children's lives and depended mostly on bureaucratic measures. The present system which is too centralized, elite-dominated and urban-oriented, will have to undergo several modifications and relate itself to the needs and convenience of various communities and young learners if India is to have a strong educational base. Its curriculum and pedagogy will have to take into account the life-view and day-to-day living style of the communities to which the children belong and adapt the learning process and its organisation to the people's needs and aspirations. The neglect of this factor of 'belongingness' of the children to the community, and its impact on learning, makes the present-day system of schooling uninteresting and even unacceptable to many.

Based on these assumptions, the Indian Institute of Education evolved an action-research project titled 'Promoting Primary and Elementary Education' for which the acronym PROPEL has been adopted. This project, conducted in 137 villages, has been eminently successful because of its cultural consciousness and community orientation. The PROPEL style of non-formal primary education has been appreciated and selected by Unesco as a 'mobilizing' 'showcase' project which demonstrates a reapplicable alternative for bringing primary education within the reach of all children, with due regard to the community life-style and people's expectations.

The culture-specific facets of the PROPEL project are: (a) *curriculum*, which emphasises (i) free scope to recite folk tales, sing traditional songs, and hold conversations about daily experiences, (ii) language and mathematics, beginning with local language and ways of calculation, leading to progressive assimilation of expected levels of learning of 'standard' language and mathematics, (iii) understanding of nature through exploration, analytical discussion, and reasoned argument, (iv) developing aesthetic sensitivity through observation, appreciation and use of colour, shape, sound, rhythm, with a view to fashioning of plastic and graphic art works in an untutored manner related to the learner's natural surroundings, (v) health and hygiene in daily life, (vi) physical and mental relaxation through simple *yogasanas*, and (vii) explorations, *with the help of the family and community elders*, in local history and geography for discovering their relevance to local conditions and to the needs of local development; (b) *class-climate* for collaborative learning through verbal and non-verbal communication by means of (i) a circular, face-to-face seating arrangement in which the instructor too is included, (ii) shared learning materials which reflect the cultural ethos of non-acquisitiveness and un-selfishness, (iii) songs and skits based on the community's environmental and cultural contexts, (iv) learning to make speeches on local subjects, and (v) group work for participatory 'peer-group' learning along with regeneration of the individualised but non-competitive, stress-free pedagogy of pre-British indigenous character.

The teachers in PROPEL are selected by the community from among community members. They are non-professionals who are willing to be trained as instructors since they wish to educate children and thus to serve the community. They are accountable to the community, and to themselves, for their performance as instructors and for proving to be like elder brothers or sisters of the young learners who need affectionate guidance. The project has ensured that every Gram Panchayat (Village Council) sets up a Village Education Committee for looking after this culture-friendly learning system and ensuring its community orientation. As to the testing of the pupils' achievements in various skills, the communities are invited to participate in the process. Pupils from several learning centres (which the children call *Apla Varg* ; *Our Class*) gather together at a central village within a walking distance of 2 or 3 kms, once in 5½ months to participate in a *Bal-Jatra* : a Children's Fair. In this air they sing, play sports and games, present skits, tell stories, and also engage in the 'game' of taking language and mathematics tests in the presence of community members. There are no passes or failures because the tests indicate to each child the next step of learning. A meal provided by the host village is shared by the instructors and pupils. Graded tests of language, numbers, and general information are supplied to groups of five or four pupils at a time, and they test themselves in public view. Thus, the project demystifies examinations and removes the confidentiality of performance. The fair provides a relaxed atmosphere and prevents examination stress.

Girls predominate in this project. They are generally drawn from the non-enrolled or dropout groups between the ages of 9 and 14. They show exceptionally good achievement in curricular studies, social skills, and understanding of the environment. They enjoy reasoning exercises, including simple experiments in science. The PROPEL project has broadly followed the principle of attending to the cultural parameters of rural primary education, especially in the case of the non-enrolled and dropout children. But the Institute is of the view that further investigations into the cultural contexts of education for rural and tribal children are essential in order to seek new paths for co-ordinating certain facets of community culture with the process of education, especially at the pre-school and primary levels, so as to bring the home and the school closer together. Also, such investigation would help delineate the ways of a meaningful fusion of essential cultural elements with innovative educational practices which could help generate the dynamism and consciousness of a wider world necessary for the people to meet the challenge of change. Thus perceived, education at the grassroots level would be an assimilative cultural movement instead of the discordant cultural scene which it currently presents.

Field-Studies

From this standpoint, the Institute made a short-term exploratory study of the cultural facets of three villages selected from three agro-climatic samples, viz., (a) drought-prone area, (b) rain-fed area, and (c) hilly tracts in the Western Ghats of Pune District in Maharashtra. In this study, interviews were held with women's groups, male and female members of households, village leaders, and other individuals active in organising or participating in local fairs, festivals and celebrations. Printed questionnaires were not used. Relaxed discussions were expected to yield both overt and covert facets of cultural ethos and practices. However, local contact persons were oriented towards the objectives of the study, manner of collecting information, recording it, and presenting it to the researchers in feedback sessions. The investigations revealed various aspects of the relationship of women with nature, their food-related rituals and celebrations, devotional practices directed towards primitive deities as well as some of the deities in the Hindu pantheon, and rites of passage from birth to maturity. The roles of men, women and children in rituals, celebrations, and festivals became clear and showed certain implications for learning activities at pre-school and primary levels.

In the three villages selected for the study, the explorations were restricted to a general view of events and practices in which children were present as witnesses and often inducted into their future roles in family and community events. Certain elements from these exploratory observations are important for providing learning activities and narrowing the gap between home and school. The pre-school and primary school organisation could be reviewed from this standpoint so as to evolve alternatives to existing practices in formal schooling in the rural areas.

Fairs and festivals in the rural areas are usually seasonal occurrences and have a clear ecological base, although some of them have acquired the garb of religious rituals. In the three villages, where climatic conditions are different and thus cause local variations in the styles of celebrations, the significance of ecological factors becomes evident. The real origins of several pseudo-ritual practices are to be found with people's age-old environmental concerns.

The case of Savardara, a drought-prone village. The soil here is rocky and vegetation is sparse. Only a few thorny trees can be seen in the otherwise barren surroundings. There

Bullock worship on 'Pala' day

Learning cane-work

Preparation of adulthood

Man serving meals to women

is great scarcity of drinking water for about eight months in a year. Drinking water is fetched by women from a well situated at a distance of about 1½ km. Each woman makes at least three trips to the well daily and brings three vessels (*handas*) full of water in each trip. For the use of domestic animals and for growing small vegetable patches, menfolk bring non-potable water from a nearby source. Both boys and girls participate in fetching water. This is not viewed as child labour. It is normal participation of children in family activities. This village, which has 96 households, has acquired some modern vehicles for transport. This has increased mobility and, in the dry season, casual labourers hire them to go elsewhere to seek employment. Three jeeps, four trucks, 20 motorbikes, 22 rickshaws, some bullock-carts and several bicycles are available. Vehicles are driven by men and occasionally by boys also. But women and girls do not drive. However, some girls ride bicycles.

The village has two major deities and a few primitive guardian goddesses (*matrikas* : mother figures). The temples of Hanuman and Bhairav (Bhairoba), who are the major deities, are the most important places for socio-religious celebrations. The 96 families in the village mainly consist of scheduled castes, Guravs (temple-caretakers), traders, potters and a few others. There is also a solitary Muslim household settled in the village over a long time. It is just one part of the community, as the various caste-groupings are.

The village celebrates all the seasonal and religious festivals. Both men and women participate in them fully. Children inevitably join in the processions and the games played by women on these occasions. The women sing special songs during the festivals. Most children know these by heart. During the naming ceremony for a newborn, and at rites of passage such as pregnancy and childbirth, children are always present, wearing festive clothes and ornaments. Both boys and girls can easily recite practically all the songs sung by women on such occasions, since during early childhood they are more with the womenfolk who look after them than with men. Along with Hanuman and Bhairav, the villagers of Savardaray respect such primitive goddesses as Janai and Jokhai, who are situated on village borders as guardian goddesses. The solitary Muslim family celebrates its festivals and rituals like *Id*, *Mohurrum* and so on. All other families participate in them and also help in them. The favourite festival of the village is the Southern New Year (*gudhi padwa*), which arrives around April as the harbinger of spring. The whole village contributes to this collective celebration. At the end of the *padwa* day, village leaders gather in the Hanuman temple, make up the accounts, and from the funds that remain unused, give assistance to individuals or families in various ways. Often, slates, clothes, etc., are given to the poorer children to attend school. There is no caste consideration.

The Misalwadi village, situated in a hilly tract, receives heavy rainfall. Its surroundings are lush with vegetation. It has 95 households. There is a Hanuman temple in the village.

The guardian of the village is the primitive goddess Malaai. Her ancient little temple is situated on the outskirts of the village, on a hill, from where she is supposed to keep an eye on the villagers and protect the boundaries of the village. A special custom in this village is that no individual goes out for work in the morning without paying obeisance to Hanuman. Interestingly, there is no priest in the village. Traditionally, a shepherd functions as the caretaker of the temple and lights the oil lamp daily. The oil for this purpose is not purchased but it is contributed by all the families in turn. Misalwadi village is a part of the village-cluster under the Kondkewadi Gram Panchayat (Village Council). From this village, no formal official of the Village Council functions because the village has always had an informally elected Council of five elders to take important decisions for the whole village. The community has continued this ancient tradition of a non-governmental Panchayat even though new Panchayat legislation has been passed by Parliament. The families in this village are committed to the custom of consensus in local government.

Misalwadi is known in its surroundings for the many festivals it celebrates. The heavy rainfall it gets provides enough vegetation nearby to enable it to celebrate tree-based festivals such as *gudi padwa* (the New Year), which requires the use of new bamboos for use as flagstaff to fly the new year banner. Bunches of fresh *neem* leaves are used to decorate the banner. This connection of festivals with forests is noteworthy. The special ritual of this day is to prepare and distribute a mixture of tender *neem* leaves, ginger powder and sugar (or jaggery) to members of the family so as to reiterate the health-giving properties of the *neem*. In every celebration, token food offerings are made to fire, water, earth, air and sky. The family can partake of the feast only after offering it to household gods, the guardian deity of the village, and a cow. An observance called *rishi panchami* (the fifth moon day sacred to *rishis*) is celebrated by all women of the village, when vegetables and cereals grown without the use of bullock-drawn ploughshare are used. This is a reminder of the way the ancient hermits and sages lived. It obliges people to protect vegetation, especially tubers, roots and leafy vegetables naturally growing on the boundaries of the village.

Tree worship is again in evidence in the practice of married women worshipping the spirit of the banyan tree on a full-moon day. The tree which is known to live for hundreds of years by continually sending out new aerial roots into the earth, is a symbol of long life. Women pray to this *vata vriksha* to give long life to their husbands so that their families may grow and live in security even as the tree does. Groups of women dress in their finery and visit the tree, which is on the outskirts of the village. Children invariably accompany them. On *makar-sankranti* day (vernal equinox), newly harvested grain, fruits and vegetables are offered to the deities and to all others invited for the celebration at the end of winter. This is truly one of the most delightful and popular harvest festivals celebrated in the rural areas.

Sweets made from sesame seeds and jaggery are exchanged with the exhortation, 'please speak sweetly the whole year long'.

On *ratha-saptami* day (seventh day of the moon for worship of the sun's chariot), the villagers worship the sun. There is much singing and dancing. On this day, a peculiar custom is observed in this village. The whole village provides a feast for all with sweets made from milk. But the responsibility for cooking is taken up by men and the women are given a full holiday from daily chores. Not only do the women have a holiday but they engage in games, sing songs, present their traditional dance, and so on. Nature related festivals like *naga-panchami* (fifth moon-day dedicated to the cobra), coconut day, when water sources are offered coconuts (on a full-moon day), and *bendur* are also celebrated. *Bendur* is meant for giving a holiday to the bullocks from farm work. Feeding bullocks with special sweets and worshipping them and also organizing bullock-cart races are special features. On *kojagiri purnima* (autumn full moon), each household contributes some milk for the community celebration. The milk is boiled by a group of volunteers in the evening, with sugar and dry fruits, for distribution to all. Children figure in all these festivals, bedecked in their best clothes, and are affectionately plied with sweets. The community leaders ensure that no child is denied this right.

The rain-fed village of Kondhanpur is quite well-to-do. Its peculiarity is that it has completely abolished untouchability and even constructed a *Buddha Mandir* just behind the temple of the community's ancient village-goddess Tukai. The Gram Panchayat of Kondhanpur gets established not by election but by consensus. Its biggest temple is that of Tukai, who is supposed to be an incarnation of the warrior goddess Durga. The village has other temples also, dedicated to Hanuman, Ganesh and Dattatreya. There are the inevitable guardian goddesses (*matrikas*), namely Kalubai and Khanjai. These primitive goddesses protect the borders of the village.

On *Dussehra* day, the whole village celebrates the festival although the families also have their own festival dinners and worship the goddess Durga at home. The village also makes a public sacrifice to the goddess. But instead of sacrificing animals, it cuts open a watermelon as a token sacrifice since its colour is red like blood. All exchange a few leaves of the *bahava* tree, which represents the exchange of gold coins practiced in ancient India. This tree is called *kanchan*, i.e., gold, in Marathi. Apart from *Dussehra*, another favourite festival is that of *Janamashtami* (Birth of Lord Krishna). On this day men hang a cradle in the local temple and do all the rituals. Men and women take out a procession. Women play traditional games in a free atmosphere without feeling shy in the presence of men. The men also play their traditional games. All children participate in the games of men and women, mostly along with parents or older siblings. On this day, all the villagers together have an afternoon feast where sweets are served. At the end of the day the senior villagers take a review of the festival and get the temples cleaned and swept for the next day's worship. On *gudi padwa* also there is a community lunch which is jointly cooked by men and women and shared by the whole community, with the children receiving all possible attention.

In all these villages, there are several features common to festive events. The Ganapati festival, which usually lasts for ten days, is the most exciting time of the year. Ganapati or Ganesh is the god of learning, of wisdom. In all rituals, he is worshipped before commencing the ritual procedure because he is the leader of the people (*gana* — people, *pati* — leader). He presides over all auspicious occasions like weddings, naming ceremonies, *pujas* of other lesser gods, and keeps a benevolent eye on everyone. Preparations for the Ganapati festival begin two to four weeks before the installation of the clay image in the home or in a public place. Musical evenings, plays, lectures by visitors, are generally arranged during this period. Children participate in this festival with great enthusiasm as Ganesh is believed to be the god of learning. Along with Ganapati his mother Gauri also is worshipped during this festival. Images of Gauri are made by tying together tender branches of wild plants specially prescribed for this ritual and protected from destruction by cattle or human carelessness. A colourful piece of cloth is wound round this green bunch and it is topped by a paper mask of the goddess. In some families a brass mask, preserved over several generations, is used. This part of the Ganapati festival is in charge of women. But both boys and girls are involved in the festival from start to finish. They particularly help collect the hibiscus flowers and lotus lilies along with certain plants which are supposed to be special favourites of Ganapati.

There are several plant-related rituals which are like fairy tales and fantasies. The wedding of the *tulasi* plant (basil) with Krishna requires that the family tend a *tulasi* plant carefully throughout the year and replant it when the old one becomes weedy. This plant has many medicinal properties and the juice of its leaves is used with turmeric powder and honey or jaggery as a mild treatment for the cough of young children. It is also used in rituals as a symbol of purity and is part of the flowers and leaves offered to all deities and especially to Lord Krishna (or Vishnu). The tasty leaves are often chewed by children and adults as a mouth freshener. The various dishes prepared during different festivals have an inherent relationship with the environment and its seasonal aspects such as plenitude of certain food items along with their nutritional and health-care significance relating to the seasons of the year. The children absorb this traditional health information without effort.

All festivals require visits to relatives and neighbours for giving and receiving presents, usually the special sweets prescribed by tradition as also clothes, new cloth, pots and pans. It is usual for women to collectively prepare some of the special food items or visit one another's houses to help. Children tag along with them and are witness to this sharing of work and products. Older girls invariably help in these tasks and gather not only cookery skills but the procedures of deciding on exchanges of presents and their social significance.

Observations

THE NATURE OF INDIGENOUS BASIC EDUCATION

The historical evidence left by British investigators reveals that the driving forces behind the indigenous system of primary education were mainly three: (a) an extension of the rites of induction of the child — especially boys — into the wider arena of skills leading to adulthood functions in the family and the community, (b) opening the doors to indigenous sources of wisdom encased in written materials and in the learning orally transmitted by the master through stories, poems and admonitions on personal and social behaviour, and (c) weaning the child away from dependence on the mother and other female caretakers at home. However, there was a cultural continuity in looking upon schooling as a phase in the process of growing up, an affirmation of the Indian view of life that it is polyphasic. Early childhood, later childhood and youth (*brahmacharya*), life of a householder (*grihasthashrama*), and retired old age away from the hurry and bustle of life (*vanaprasthashrama*) are phases requiring a clear-cut set of duties and human relationships. Indigenous child education had a strong moral and social goal. It had only a marginal economic goal within its largely agrarian setting. It was also a part of the ritualistic behaviour of the family to which the child belonged and, therefore, an indivisible aspect of local culture.

THE ANGLO-SAXON SCHOOL IN INDIA

The openly stated objectives of the school system established in India around 1765, and later confirmed by the imperial government, were to (a) wean away the Indian people from their culture and shape them into 'brown Englishmen', and (b) provide subordinate personnel for revenue collection, judicial posts, and train clerical cadres for the convenience of British administration in India. So as to ensure that sedition may not raise its head and for facility of supervision and management by educational administrators, the curriculum and text-books were prepared by government. Uniformity was insisted upon, reinforced by public examinations even at grade IV and VII levels. These examinations were used as 'sorting machines' for eliminating as many learners as possible so as to keep the numbers of certificate holders small enough to correspond with the available government jobs and also to keep the educational expenditure of government to the minimum essential for its administrative needs.

The driving force behind this imperially prescribed system was the subordination of natives and stabilisation of colonial rule. English was therefore made the medium of instruction from grade V onwards for training for relatively higher posts, while the vernacular medium was permitted for the posts of village level functionaries. For the English medium school course, the matriculation examination was held by universities. It applied further brakes on the numbers of 'qualified' persons by being extremely strict and generally failing nearly 75 per cent of the candidates. This system thus intentionally kept education from spreading, both through the destruction of the indigenous system and through the construction of a system which functioned as a sorting machine for keeping the number of successful candidates limited. The Anglo-Saxon system brought by the imperial government frankly stressed the 'education of the few' and the non-Indianization of those few who went through the system.

Looking at the educational situation on the eve of Independence, it became obvious that the imperial government had eminently succeeded in subjugating Indians not simply during its direct rule over them but for a much longer period. The gap between the culture of India's educated elite and that of the rural masses persists and has gone on widening. It persists, and the Indian mind is now confused. The question is whether this confusion can be overcome without reverting totally to the cultural past, which has now become irrelevant, or accepting the ways of the West which also are equally irrelevant to India's development needs and goals. One feature of the imperial style of government that has continued to prevail with detriment to the people's values, aspirations and freedom to decide their educational system, is bureaucratic patternalism and its attachment to the principle of centralised control. This is seen in every sphere of administration and especially so in education, where even the village school teacher, being a government servant, is steeped in the bureaucratic spirit. Teacher-training curricula, decided more by government officers than by thoughtful educationists, promote this bureaucratic linkage from the top to the bottom. If the people's culture is to assert itself in deciding the various phases of the education of Indian children, this situation would have to be critically researched for evolving remedies to release rural child education and primary education from its Anglo-Saxon bondage.

VILLAGE-STUDIES : EDUCATIONAL REFLECTIONS

(a) *Cultural identity* : The explorations in three villages situated in climatically different tracts have brought out several aspects of culture and modalities of child socialization. Arising from human interaction with nature and a time-honoured value system which transcends the day-to-day subsistence concerns of the rural people, the cultural commitments of the family and community shape the child's personality almost from the moment of birth. However, the local primary school transplanted by the state in this cultural milieu remains unaware of the education received by the child through participation in various rites, rituals, celebrations, life-style and allocated tasks. It is this education that shapes the child's personality and makes it feel secure within its cultural setting. It knows who it is, what it is expected to do, in what way, and how to relate itself to the kinship structure and the neighbourhood. It learns a tremendous amount about nature and develops emotional bonds with its different seasonal manifestations. All this knowledge, gathered at first hand, infuses self-confidence in the child and forms its cultural identity.

(b) *The school and the child* : In the process of growing up the child is helped by the various stake holders, parents, kin, neighbours and the community as a whole. The school teacher, being an outsider, is not a participant in this process of incidental but fairly well-organised cultural learning. For the rural child, therefore, the teacher in the formal school is a stranger and the school is a place where it has no protection from parents, older siblings, or relations and neighbours. School entry for the child is traumatic, an entry into insecurity, into a world of which the ways are quite different from what the child has experienced in its cultural milieu. This is often the reason for non-enrolment of rural children and of a large dropout rate. Not only the child but the parents also are distanced from the school, since it is only the teacher who is designated by the system to transmit education according to a state-prescribed curriculum. The textbook is the only tool of education, but the standardised, formal language it uses is far removed from the local vernacular. Its lessons deal with subjects with which the child has no cultural familiarity. The so-called co-curricular activities are also prescribed from a 'national' and 'international' standpoint about which the community, parents and children know next to nothing. Local festivals and celebrations receive a small place in the annual list of state-prescribed holidays, but the school does not figure in them as a cultural participant. The parents do not know what the curriculum intends to do nor what the textbook teaches. As a result, this education can receive hardly any support at home. The results of school learning fail to correspond with the parents' expectations of schooling and they withdraw their children from school. The children, by and large, do not regret this withdrawal and happily return to the security of the culture imbibed by them before school entry. This picture was seen in the three villages under investigation. It reinforced the overall research findings in this respect.

(c) *Men and women* : All the three villages gave evidence of gender-related cultural practices at home, at the workplace outside the home, and in community life. While the work of women relates to organising the rituals and celebrations within the home and family, men take up organisational tasks outside at the community level. These activities are interdependent and creativity is seen in the tasks performed by women and girls within the home and by men and boys outside the home. For festival processions and religious gatherings, all adults and children come together, transcending gender and caste barriers. Even religions are no bar to such celebrations as seen in the *Moharrum* festival at Savardaray and in the construction of the Buddha temple at Kondhanpur.

Women and girls delight in excelling in cookery and decorating the home and its yard at festival time. Men are enthusiastic about putting up a decorated *gudhi padwa* flag or decorating the bullocks for the *bendur* festival. However, on a certain day in the year, the men do the cooking and serve the women, thus reversing gender roles. On the occasion of the *kojagiri purnima*, the men undertake to boil and distribute sweetened milk, and they do the same on *ratha-saptami*, dedicated to the sun's return to the north.

(d) *Relationship with environment* : On festival days or during religious rites, token food offerings are first given to the earth, the sky, water, fire and air, since without a combination of these elements life cannot be sustained. Certain festivals stress the greater closeness of women to nature. Worship of the banyan tree, caring for the plant and gathering naturally growing vegetables denote this close relationship and have a symbolic significance for the well-being of the family. Also, the use of bamboo and *neem* leaves on *gudhi padwa*, distribution of *kanchan* leaves on *Dussehra*, offering of hibiscus flowers and lotus flowers to Ganapati, using mango leaves and marigold flowers for garlanding the door-frame of the house on *Dussehra* and *Diwali* days, demonstrate the delight people take in their natural surroundings where trees and other vegetation are still to be found. Preserving this natural wealth, appreciating the role even of the cobra in these rural surroundings, is part of the child's cultural education. In this culture, man and nature are not adversaries but friends. Man does not conquer nature but lives with it in a relationship of give and take. These practices imply a very important philosophical stand on the relationship between man and nature common to eastern civilisations, where nature has generally been quite bountiful.

(e) *Economic transactions* : All the three communities under investigation had a tradition of economic transactions which cannot be monetised. The local economy is not controlled by bureaucratic procedures nor does it figure in the statistics of state bureaus of economics. For instance, the surplus festival funds distributed by village elders to needy persons are part of the community's welfare activities and do not figure in government expenditure on welfare. Some festivals appear to generate temporary contract jobs for erecting *pandals*, making leaf plates and bowls, collecting firewood, fetching large quantities of water, and so on. The milk contributed for the *ratha-saptami* festival is non-monetary. Most community meals have large contributions in kind which no one bothers to put into account books. Worshippers of the goddess Malaai maintain the temple and contribute oil for the sacred lamp all the year round. Barter of various commodities seems fairly common among neighbours. In weddings and other ceremonies, exchange of clothes and new pieces of cloth is a confirmed practice in which the 'giving well' aspect is stressed more than how much or what one might receive in exchange. Local vegetation is often used by everyone for some purpose or other but is not vandalised, so that it remains a community asset. Thrift is practised because the saints have exhorted the people to do so; but no visitor who may drop in at mealtime is allowed to go without partaking of some food. This is not calculated in terms of money.

Children in the rural areas get habituated to transactions of this kind. In the school text-books nothing of this nature is reflected. The community is often asked by educational bureaucrats to construct school-rooms or give equipment to the school. But the immediate users of primary education, i.e., the parents and the community, are never asked whether the curriculum, pedagogy, or economic aims of schooling correspond with their expectations and whether they could help make education culturally relevant. The gap between the economic concepts acquired by the children in the community and those adopted by the state in a bureaucratised school system being seriously at variance, the cultural confusion of the learning process gets exacerbated. The aim of Westernized basic education being employment, and that of the cultural education of the children being collaborative interdependence for the day-to-day business of living and growing, a clash between the two becomes inevitable.

Towards a Systemic Change in Education

This section cannot resist quoting Gandhiji at the outset as the background for its arguments:

> Nothing can be further from my thought than that we should become exclusive or erect barriers. But I do respectfully contend that an appreciation of other cultures can fitly follow, never precede, an appreciation and assimilation of our own. It is my firm opinion that no culture has treasures so rich as ours has. We have not known it, we have been made even to deprecate its study and depreciate its value. We have almost ceased to live it. An academic grasp without practice behind it is like an embalmed corpse, perhaps lovely to look at but nothing to inspire or ennoble. My religion forbids me to belittle or disregard other cultures, as it insists under pain of civil suicide upon imbibing and living my own.

The validity of the Western model of schooling has been challenged not only by Gandhiji but in all the countries which were colonised, whether in the east or in the south. This model had ridiculed and devalued community-based systems of education with their culture compatible organisation. Diversity was rejected. The community's ways of socialisation were condemned as primitive even in a country like India which has evolved knowledge in its many branches over thousands of years. Western science and technology and Western economics were propagated as elements of modernisation. The wheel of educational thinking has, however, been turning full circle and academics are pointing out that the foundational education of children cannot ignore its cultural contexts, that it can be successful only when the parents and the community participate in fashioning its aims, content, pedagogy and organisation. Non-formal education, alternatives to the present unsuccessful schooling system, and going beyond Western modernism to post-modern thinking which insists that the hegemony of the state over education must end, are some of the indicators of the commencement of an educational renaissance in Third World societies. The enabling of the stake holders in education to assert their viewpoint on the fashioning of the future of society can be envisioned as the core of the educational scene in the post-modern era.

The configuration of education, particularly foundational education ('basic education' in the recent terminology of international donor organisations) needs to consist of elements which are local and culture-friendly. At the same time, widening of the learners' horizons should be possible through multimedia programmes, leaving their use not to state prescription but to the learners with their various goals and interests. Such widening of horizons is essential not only for ex-colonies like India but for the ex-colonisers as well. There could be programmes having diversified curricular offerings, taking the learner from the local level to the regional, national and even international levels of knowledge acquisition through a process of life-long learning. It is becoming clear that the very concept of school has to change, whether Western or Eastern. But whatever may be the learning opportunities opening out henceforth, they need to be shaped and used by people everywhere from their own cultural moorings, in the context of their integrated civilisational view. There would then be educational diversity which is essential for overcoming the control of the state and for handing back to the people the future of their cultures and their children's destinies.

But this is a rather vague vision at the present moment. For giving it sharpness, extensive studies in the cultural contexts of education need to be undertaken in diverse cultures.

References

[The authors acknowledge the guidance of Dr. Chitra Naik]

Carnoy, M and Samioff, J., *Education and Social Transition in the Third World* (Princeton University Press, 1990).

Cremin, Lawrence, *Toward an Ecology of Education* (Excerpt from *Public Education*, John Dewey Society, Pub:USIS, New Delhi, 1976).

Dharampal, *The Beautiful Tree : Indigenous Indian Education in the Eighteenth Century* (Biblia Impex, Pvt. Ltd. New Delhi, 1983).

Di Bona, *One Teacher, One School : Adam's Reports* (Biblia Impex, Pvt. Ltd. 1983).

Gandhi, M.K., *India of my Dreams* (Navajivan Publishing House, 1995).

Iyer, R. N., *The Essential Writings of Mahatma Gandhi* (OUP, 1995).

Joshi, P.C., *Cultural Communication and Social Change* (Vikas Publishing House, New Delhi, 1992).

Naik, J.P. and Nurullah, S., *A Students' History of Education in India* (McMillan & Co., Bombay, 1975).

Takala, Thomas (Ed.), *Quality of Education in the Context of Culture in Developing Countries* (Uni. of Tampere, Deptt. of Education, Tampere, Finland, 1994).

4

The Bose Foundation School

Baidyanath Saraswati, Shivashankar Dube & Ram Lakhan Maurya

Swaraj in Education

Baidyanath Saraswati

THE eighteenth-century European colonization and Christianization indicated the direction in which the wind of education was to blow in India. But this is not a 'thing' to celebrate. It is common knowledge that the westerly direction of that wind has remained unchanged well into the twentieth century. We do not know why the wind is so particularly strong, but we do know that there are other directions in space-time.

Among early colonialists, there were liberal-minded Europeans who felt the desirability of "preserving the ancient culture of India from the state of rapid decay into which it [had] fallen on account of the loss of royal patronage". They also saw the utility of the careful study of ancient Indian literature by Western scholars and the adoption of English as a medium of instruction for the Indian people. The missionaries and their friends, on the other hand, looked at Indian culture with utter contempt. They believed that Western 'light and knowledge' should take the place of Eastern culture and religion. The foremost among them was Charles Grant, the father of colonial education in India, who painted an exaggerated picture of the 'depraved' condition of Indian society. He analysed its cause and suggested a remedy: "The causes of the miserable condition of the Indian people were ignorance and want of a proper religion. The situation could only be improved if Indians were educated through the English language and finally converted to Christianity." Lord Macaulay, a torch-bearer in the path of colonial progress, also recommended the spread of Western learning through the medium of the English language. In his infamous minutes he wrote, "a single shelf of a good European library [was] worth the whole native literature of India and Arabia". He talked of creating "a class of persons who would be Indian in blood and colour but English in tastes, in opinions, and in intellects".

What flowed out of these two streams of thinking was a pattern of education with the ultimate objectives of (a) securing servants for public administration; (b) diverting young men and women from the study of oriental to occidental literature; and (c) arousing in young hearts a passion for Western knowledge and culture. The objectives were carried out by (i) the creation of a government department of public instruction; (ii) the establishment of universities, colleges and graded schools; (iii) the training of teachers; (iv) the introduction of government grants-in-aid; (v) the maintenance of a few educational institutions under the direct control of the government and allowing private educational enterprise by missionaries and non-officials; and (vi) the recognition of educational institutions by the government department and universities.

At the culmination of the colonial process India inherited: (a) employment-oriented education; (b) Westernization of the content of education; (c) public examinations so used as to impose uniform curricula and textbooks; (d) a class of persons educated in a foreign language; (e) neglect of indigenous systems of education; and (f) the withdrawal of religious education through direct educational enterprise.

The Indian National Congress was formed in 1885. The *Swadeshi* movement brought about a great ferment of educational thought. It demanded Indian control of Indian educational policies, teaching love of the motherland, no servile imitation of England, and the removal of the domination of English. Unfortunately, hopes of educational reconstruction with a bolder and freer hand have not yet materialized.

India struggled for freedom, and it did achieve political independence. It asked for Indian control of Indian educational policies, and it got it. It wanted to teach love of the motherland, and it is doing so. But is independent India free from servile imitation of the West? Has the domination of English gone? Has the Western system of education enriched Indian culture? Has it added at all to its happiness? Does India need such education?

India debated Gandhi's great idea of education for nearly a century. It seems to be in earnest in its endeavour to educate people. More than 90 per cent of the country's rural areas now have schooling facilities within a radius of one kilometre. The national policy on education is revised periodically and the investment on education has now reached 6 per cent of the national income. Yet the picture is unclear.

The national policy aims at (a) promoting national progress; (b) creating a sense of common citizenship and culture; (c) strengthening national integration; and (d) giving greater attention to Western science and technology.

The national organization of education is formed by (a) sharing of responsibility between the governments at the centre and in the states; (b) administering at the national level, state level, district level, and local level; (c) establishing the Indian Educational Service as an all-India service to bring a national perspective to education; (d) forming education tribunals fashioned after administrative tribunals; and (e) giving shape to the national system of education through such institutions as the University Grants Commission, All-India Council for Technical Education, Indian Council of Agricultural Research, Indian Medical Council, National Council of Educational Research and Training, National Institute of Educational Planning and Administration, National Council of Teacher Education, National Council of Adult Education, etc.

It is not surprising that the results have been devastating. National policy has been influenced by the political expediency of conciliating the people and education has been politicized. There is greater and greater dependence on the government bureaucracy. Uniformity has been imposed through common education structure, common school system, common curricular framework, and universal literacy. Deculturation has come about through religious neutrality or secular education. Employment-oriented education has led to the privatization and commercialization of learning, and a colonialism of the English educated urban elites has come into being.

Is there a way out of this slavish system? 'Where there is a will there is a way' is the old proverb. India's 'national policy' on education is at the moment geared to horizontal (westward) movement in time. If education is meant to provide an ever greater degree of moral and mental sophistication, it has to develop through vertical movement. Exploration of the vertical dimension with free will and accompanying intellectual responsibility implies a vertical movement from the past to the future. This essay is an invitation to appropriate for ourselves a vertical movement in education. To do that let us go to the sacred city of Kashi on the Ganga, the source from which India's wisdom tradition sprang.

Education in Kashi

Three cities, Rome, Mecca and Kashi (Varanasi or Banaras, as it is also known) need no introduction. So far as the continuity of the classical tradition is concerned, Kashi, the city of light surpasses all the civilizational centres of the world. And yet it is a puzzling city, a city which has an infinite capacity to absorb the most beauteous and bear the most repulsive things. It is like Shiva the Nilakantha, the supreme deity of the city.

Several centuries before Christ, Ajatsatru, the king of Kashi, defeated the Brahman Gargya in the *shastrartha* (debate on scriptural matters), and Gautam the Buddha turned here in Isipatan (modern Sarnath) the wheel of his *dharma*. In course of time Kashi grew into a great seat of learning, surpassing Takshashila and all other educational centres of India. Huien-Tsang was struck by the scholarship and devotion of the Brahman students and the Jain and Buddhist ascetics of Kashi in the seventh century. Earlier, in the fifth century, Fa-Hein noted that "here in this great city there were thirty monasteries and about three hundred Buddhist priests, and the Hindus had about one hundred temples with ten thousand sectaries and their principal God was Maheshwara whose copper image was a hundred feet high". The ancient travelogues also indicate that Brahman, Jain and Buddhist scholars lived in harmony while engaged in their pursuits of learning.

In later times Kashi gained a high reputation and attracted scholars from far and near. It became famous for the assembly of the *pandits*, which organized *shastrartha* on disputed matters of social importance. Any decision arrived at by the learned assembly of Kashi was accepted as the norm by the entire Hindu community. The chief source of the Banaras School of Law, one of the five recognized schools of Hindu law, was Vijnaneshvara, the author of the *Mitakshara* of the twelfth century. Bernier described in detail the methods of study of the *pandits* of Kashi in 1667. When Ward visited Kashi in 1917 he found forty-eight teachers instructing 893 in the *Vedas* alone, and seventeen teaching 218 disciples the mysteries of Panini's grammar.

The lamp of Sanskrit learning was kept alive in Kashi for a long time, particularly by the Maharashtra and Kannada Brahman families who migrated to the city at the beginning of the sixteenth century. These families remained at the helm of Sanskrit scholarship for no less than three centuries; later the *pandits* of Mithila and Bengal came to their support.

Besides being the leading centre for the study of religion, philosophy, medical sciences and astrology, Kashi enjoyed a reputation for its handicrafts and commerce even during the pre-Buddhist period. Banarasi silk fabrics were exported to all parts of India. It was famous for perfumes, scented oil, ivory works, and sculpture. Its contributions to Indian vocal music are the melodious *thumri*, *dadra* and *tappa*.

Education in ancient Kashi, as elsewhere in the India of that time, was a self-organizing system. Generally speaking, the schools had no buildings of their own. Temples, private buildings donated by pious men, the houses of teachers, and even the *ghats* on the Ganga, were the glorious centres of learning. Of the teachers, the majority were Brahmans who taught more through a sense of righteousness rather than by consideration of economic gains. The pupils did not race through examinations to pick up lucrative jobs. As the outcome of a humanistic aspiration, education was considered at that time a life-long pursuit. Freedom in academic life was so firm that even the strongest ruler could not tamper with education. Education in the arts was largely informal and orally transmitted within hereditary, non-competitive, monopolistic, endogamous groups. However, this self-organizing system declined during the long period of changing political patterns.

The medieval rulers remained indifferent to India's integrated religious and metaphysical system of education. In the beginning, like their *pandit* brethren, the *ulemas* also enjoyed intellectual freedom. But as Islam's earlier democracy was replaced by authoritarianism, Muslim education became dogmatic and inward-looking. Hindus and Muslims built separate centres of learning. However, as Persian became the official language of administration and justice, many Hindus had to learn Persian by force of circumstance.

And yet Kashi provided sufficient incentives to all those who made it their home: artists, craftsmen, philosophers, traders and ordinary persons, regardless of their creed and caste. It did not permit a sectarian outlook to prevail upon the cultivation of excellence. Or else Babu Miyan, a Muslim, would not have been a specialist in the Hindu *shilpashastra*; Samsuddin, another Muslim, would not have been a *pandit* in Hindu astrology; Ulfatbai, a Muslim lady, would not have made an endowment for the playing of the *shahanai* by Muslim musicians in the temple of Vishwanath; and Muslim weavers would not have made auspicious wedding garments for Hindu ladies.

Kashi holds people by generating a deep sense of attachment, which is not restricted to Hindus alone. The illustrious Iranian poet Sheikh Ali Hazeen did not like to leave Kashi for anything, and wanted that even after his death his body should lie in this holy city of light where every scholar is treated like gods Rama and Lakshmana:

> *az banaras na ravam, mabadi-am ast een ja;*
> *har barahman-pisaraya lachhaman-o-ram asta een ja.*

The Sheikh further noted: "The people of Banaras admired and respected me and my talents in the like manner and that is what I longed for during all these roamings from Ispahan, my native place, to Banaras. Hence I feel satisfied to remain in Banaras till death". He died in 1180 and was buried in the Fatman of Banaras, as desired by him.

European colonization initiated the final death blow to the self-organizing system of Kashi's intellectual tradition. With the establishment of the Kashi Raj in 1725 and the control of the East India Company in 1757, a new educational system began to grow. Initially the British Company was reluctant to take responsibility for education. But later, in 1813, it felt that "it must educate the sons of influential Indians for higher posts under the government and thereby win the confidence of the upper classes and consolidate its rule in India".

The Banaras Government Sanskrit College owed its establishment to this political consideration. It was founded in 1791 by Jonathan Duncan, Resident of Banaras. In 1813, a Persian class was started in this Sanskrit college to teach those students of Hindu law who wished to be appointed *pandits* in the British courts; but not a single *pandit* ever availed of this opportunity. So in 1833 the Persian class was temporarily suspended. In 1841 it was begun again under the orders of the Governor-General; but the situation remained the same, and in 1844 the Persian class was finally removed to the English College, which had been started in 1830 under the title of the Banaras English Seminary or the Banaras Government School. Jayanarain Ghosal, a native Bengali, established here the first English school in 1817, and opened the door for English education. Christian missionaries began to operate through educational programmes with a view to capturing the intelligentsia of Kashi. This led to the establishment of the Church Missionary Society in 1888.

From enslavement and dispossession sprang a new consciousness, supported by large-hearted Europeans. In 1897, Mrs Annie Besant founded the Hindu College and established the Theosophical Society to "revitalize the faith in Upanisadic Hinduism among those who were enamoured of English education and Western culture". While Sanskrit education in the Government College was christianizing and transforming and the British administration was patronizing the *pandits*, a number of scholars became interested in the development of Hindi. The Nagri Pracharani Sabha was founded in 1893.

Towards the powerful wave of Gandhi's freedom movement the intellectuals of Kashi were divided into two groups. Most *pandits* opposed Gandhi's attitude towards Brahmanic orthodoxy; others wholeheartedly supported his liberalism and nationalism. The establishment of the Bharat Dharma Mahamandal in 1902 and the foundation of the Syadbad Mahavidyala in 1905 were due to the response of the orthodox tradition. Mahamana Pandit Madan Mohan Malaviya, a liberal Brahman who wanted to instil the spirit of nationalism among the youth of the country through modern educational system, established the Banaras Hindu University in 1909. Further, the Mahamana wanted students to be firmly entrenched in the glorious tradition of this country, which obviously meant to him the Sanatani Varnashrama Dharma — the eternal order of human life visualized in the Hindu tradition. The nationalists, who did not approve of his attitude toward the *Varnashrama* order, established yet another educational institution. In 1920 the Kashi Vidyapeetha was founded by Mahatma Gandhi and began to operate as a national school for revolutionaries and freedom-fighters. Those who joined the Vidyapeetha were largely anti-Brahmans. The clash of ideologies came to the fore when the nationalists enrolled members of non-Brahman castes and Muslims for a course leading to a diploma called 'Shastri', a title derived from the Sanskrit system of learning, traditionally held by Brahmans. In imitation of the modern university system, the Vidyapeetha, having given up the title of 'Shastri', now awards bachelor's and master's degrees.

Although Kashi today is an important city enjoying all the benefits of modernization that are normally available to any other city of this size and resources in India, it has a unique way of modernizing its tradition. This is reflected in the formation and functioning of its three universities and three temples of Vishwanath — the presiding God of this sacred city — each symbolizing a cultural type. The Varanaseya Sanskrit University and the new Vishwanath temple, founded by Swami Karptrijee, represent the orthodox Brahmanic tradition; the Kashi Vidyapeetha and the golden temple of Vishwanath, managed by the state government, represent the liberal tradition; and the Banaras Hindu University with its magnificent temple of God Vishwanath and teaching of modern science and technology, represents the most modern view of tradition. However, the common factor which binds all these three cultural types together is the unshakable trust in tradition.

The experience of an experiment may perhaps be best expressed through the analogy of drama. As previously indicated, the context of education presents itself at the most magnificent theatre of Kashi; its text draws upon a humanistic appreciation of tradition having a universal significance; its melodies and notes produce a new life and freedom; and its joys proceed from truth and goodness.

The Sutradhar of an Experiment

On 1 November 1972 a Foundation was formed in Kashi in memory of Professor Nirmal Kumar Bose, the famous Gandhian anthropologist who had the fortune of serving Gandhi for several months in Noakhali in 1946-47 as his Bengali interpreter and private secretary. The Bose Foundation started with the objectives of promoting and propagating such elements of cultural tradition as are helpful to the development of Indian society in particular and to the cultivation and progress of a peaceful and loving coexistence of human societies in general, carrying on both applied and fundamental research in cultural anthropology, evolving an appropriate methodology for the studies of complex societies, and training researchers in the field of cultural anthropology. To begin with, the Foundation devoted considerable effort to examining the various elements of Kashi's cultural tradition, namely ascetics, *pandits*, temples, rituals and pilgrimage. This was followed by applied anthropological research on the status of widows and the problem of indebtedness among the scavengers of Kashi.

In October 1978 the Bose Foundation was made responsible for the Annapurna Shikshalaya, a social welfare trust located in Kashi at Gauriganj. The person who entrusted the Foundation with this responsibility was the late Professor Asit Bhattacharya, a student of Nirmal Bose and a great-grandson of Sarojini Devi who had founded the Shikshalaya. The Foundation felt that this good fortune of having been invited to perform a certain role in social work would not have come without the grace of *guru* and God. Nirmal Bose, in whose memory it was formed, was always concerned with making human beings. He was an ally in the struggle for freedom and his true terms of reference were the poor. In 1930 he organized a Khadi Sangha in a slum at the outskirts of Bolpur town located about 2 km away from Rabindranath Tagore's Shantiniketan and Visva Bharati. He set up there Shikhagar, a night school for adults of the poor 'untouchables' — the Muchi, Hadi and Bauri castes. Following Bose's ideology and experiment in social reconstruction, the Foundation found it fit to accept the new responsibility.

When the Annapurna Shikshalaya was handed over to the Foundation, it was in a moribund state with two widows, a lifeless primary school for children, and a loom for weaving carpets from tattered clothes. To streamline its activities, the Foundation reorganized itself into three departments. The old school was named Sarojini Vidyakendra after the name of the founder of the Shikshalaya, and the craft centre was called Kuntala Shilpakendra, after Kuntala Devi, the eldest grand-daughter of Sarojini Devi and the mother of Asit Bhattacharya. The school and the craft centre were placed under the charge of Shiva Shankar Dube and Ram Lakhan Maurya, two men of glorious light, the *sutradhars* who were earlier involved in the study of widows and scavengers of Kashi. The Bose Research Centre continued its academic activities as before.

The Theatre

In its early enterprise the Foundation had opened the whole panorama of Kashi. With the eye of a team of researchers it had seen a light within Kashi: a city of cultural pluralism, a city of *bhoga* and *moksha* — materiality and spirituality — a city of affluence and grinding poverty, a city with two major religions and three cultures, and a city where the sacred was in tune with the secular. The studies of widows and scavengers let it traverse the area of darkness. The Foundation was now called upon to rekindle a 'power for better things'.

Gauriganj is a magnificent complex of multiple cultures, sheltering at least thirty-seven Hindu castes and fifteen Muslim endogamous groups. Besides the first three *varnas* — Brahman, Kshatriya and Vaishya — there are castes of silk weavers, dyers of clothes, potters, ironsmiths, goldsmiths, tinkers and braziers, barbers, grocers, fowlers and hunters, palanquin-bearers, milkmen, washermen, cultivators and sellers of green vegetables, tavern-keepers and wine-merchants, manufacturers of salt, cobblers, *shahanai* players, and those who prepare and sell cups made of leaves. Most of these caste groups have now given up their traditional callings. There are only a few wealthy individuals; all others live in abject poverty.

The 'hub' of Gauriganj, a riot-prone area, is surrounded by a number of culturally significant shrines (Map 1) such as the cremation *ghat*, where in the Age of Truth (*satayuga*) the truthful king Harishchandra had to serve the master of the crematorium; the Shivala Imambara, where Muharram, a Muslim occasion of sorrow, is observed; the Assi *ghat*, where the *lilas* of Rama and Krishna are performed; the Tulasi *ghat*, where the great saint Tulasidas composed his celebrated *Ramacharitamanas*; the Durgakunda, where fairs are held in honour of the Mother Goddess; the Krimikunda, the pond of the worms, where healing and fertility rites are performed; the Lolarkakunda, where the 'Trembling Sun' is worshipped, and several other springs associated with legends and myths of therapeutical value. Here are located countless shrines of gods, the birthplace of the Jain *tirthankaras*, Sikh gurudwaras, mosques and tombs, cathedrals and missions, and a very large number of *mathas* and *akharas* of ascetics.

This southern sector of the city, according to the *Padma Purana*, is Kedara Khanda, the third segment of Kashikshetra. In the popular oral tradition the southern zone is called Shiva Kashi. In the human domain it is a meeting ground for Hindus and Muslims, north Indians and south Indians, ascetics and *pandits*, rich and poor, rural and urban, traditional and modern, and so on. This is the area where, for the first time, modern educational institutions were founded — the first English school, the first Hindu college, the first modern university, the first theological school for Jainism, and the first Arabic university. It is also the breeding ground for indigenous *pathashalas*, *madarassahs* and private modern schools.

Making an experiment in education in such a magnificent theatre of culture was a test of intellectual powers and patience. We had both advantages and disadvantages. The greatest advantage was easy access to the visual text of a living cultural tradition with its constant newness. The disadvantage was of facing iconoclasts of the Indian tradition who dogmatically affirm Western education. Though in thought this kind of challenge may appear destructive, in reality it is constructive. It makes differences between normal and abnormal conditions in education explainable; it provides for enduring contemplation; and it ultimately leads all activities to the carried end.

It is with this urge that the old school at the Annapurna Shikshalaya was directed to a new life and new rhyme. The new venture was envisioned as a 'lab school' alive with the ideas of making experiments with *swaraj* in education, developing aesthetic sensibility and cross-cultural understanding, evolving new perspectives and new methods in primary education, and making the school a self-organizing, self-supporting, non-commercial, non-governmental institution.

VĀRĀNASĪ

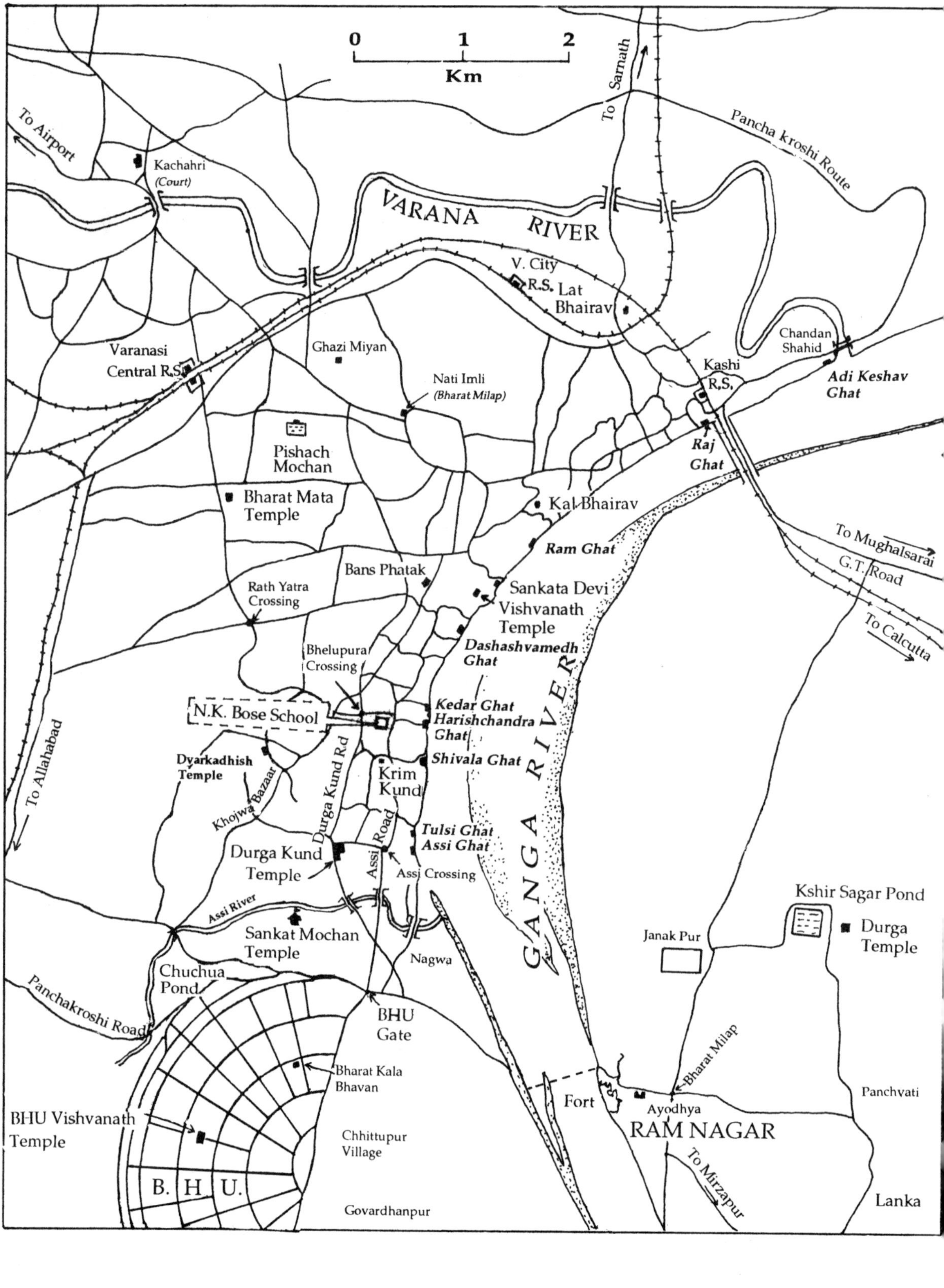

0 1 2
Km
To Airport
To Sarnath
Pancha kroshi Route
Kachahri
(Court)
VARANA RIVER
V. City R.S.
Lat Bhairav
Varanasi Central R.S.
Ghazi Miyan
Nati Imli
(Bharat Milap)
Chandan Shahid
Kashi R.S.
Adi Keshav Ghat
Raj Ghat
Pishach Mochan
Bharat Mata Temple
Kal Bhairav
Ram Ghat
To Mughalsarai
G.T. Road
To Calcutta
Bans Phatak
Rath Yatra Crossing
Sankata Devi
Vishvanath Temple
Dashashvamedh Ghat
Bhelupura Crossing
N.K. Bose School
Kedar Ghat
Harishchandra Ghat
Shivala Ghat
Dvarkadhish Temple
Krim Kund
Durga Kund Rd
Khojwa Bazaar
To Allahabad
Tulsi Ghat
Assi Ghat
Assi Road
Durga Kund Temple
Assi Crossing
GANGA RIVER
Assi River
Sankat Mochan Temple
Nagwa
Kshir Sagar Pond
Durga Temple
Janak Pur
Chuchua Pond
Panchakroshi Road
BHU Gate
Bharat Kala Bhavan
Bharat Milap
Fort
Ayodhya
Panchvati
RAM NAGAR
BHU Vishvanath Temple
Chhittupur Village
To Mirzapur
B. H. U.
Govardhanpur
Lanka

Assi & Neighbouring Muhalla

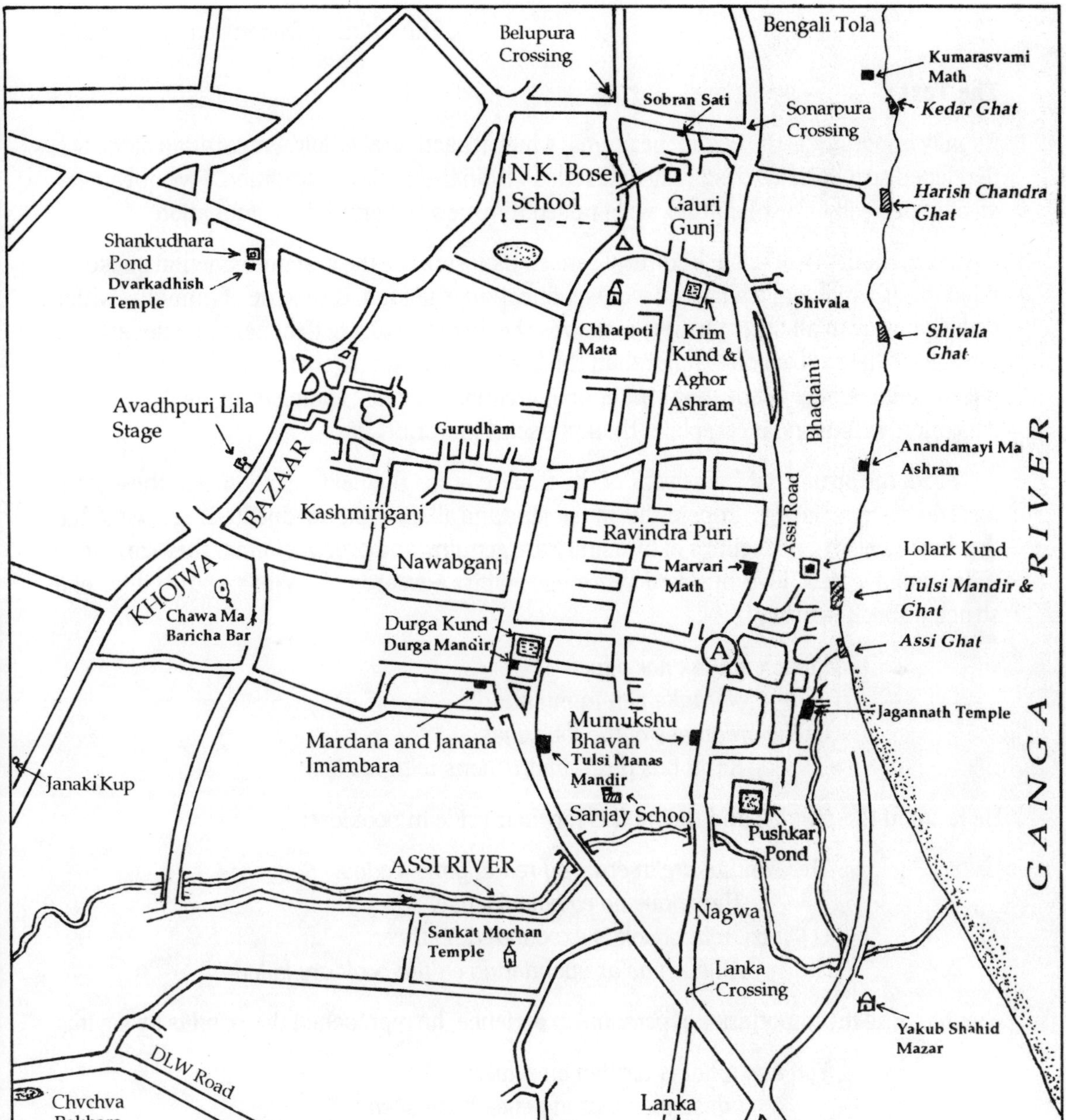

The Text

Equally important is the text of the drama, which its activities follow. Preparation of the text developed sequentially, designing new styles within the circle of a tradition. Five questions, all of which call for explanation, were raised to review the grounds of education.

From the depth of Kashi's cultural consciousness arose the first fundamental question. What is education? In traditional words, education is *vinaya*, the virtue of humility, which is the gateway to all other virtues; *vinaya* is the attribute of a perfect person; education is the light of the soul which is other than the body; a place where that light does not shine is not an educational institution; 'interior education' is real, complete in itself; 'exterior education' is illusory, incomplete, though useful in worldly life.

From the structural framework of the theatre arose the next question. Are these non-literate weavers, potters, ironsmiths, musicians and all the rest, uneducated? Knowing that their works of art and feelings of humanity are genuine and overwhelming, they cannot be called uneducated. Five hundred years ago Kabir, a weaver of Kashi's spirituality, had struck at book learning:

I touch not ink nor paper
Nor take pen in my hand;
Of the greatness of the four ages
Kabir has given instructions with his lips.

He taunted the *pandits* and the *mullas* for their pride in booklore:

The Pandits are in error by reading the Vedas.
They have no common sense.
O Quazi, what book is expounded by thee;
All such as are pondering on the book are killed.

Emphasizing the importance of personal experience, he reproached the *pandits* by saying:

You say what is written on paper;
I describe what my eyes have seen.

For, he believed:

Should all the earth be turned into paper
and all the trees into pens;
Should the seven seas be turned into ink,
yet could not an account of God be written.

ively children

Be cheerful

Be patient

Together in love

Training body and mind

Meditation of peace inspires calm

Yogasana, a perfect performance

Physical education

Also for Gandhi (1908), literacy alone was not enough:

> It (education) simply means a knowledge of letters. It is merely an instrument, and an instrument may be well used or abused. The same instrument that may be used to cure a patient may be used to take his life, and so may a knowledge of letters. We daily observe that many men abuse it and very few make good use of it; and if this is a correct statement, we have proved that more harm has been done by it than good . . . . I do not for one moment believe that my life would have been wasted had I not received higher or lower education . . . . I have not run down a knowledge of letters in all circumstances. All I have now shown is that we must not make of it a fetish . . . . Our ancient school system is enough. Character-building has the first place in it and that is primary education. A building erected on that foundation will last.

From the realization of the life-view of Kabir and Gandhi the third question arose. Can the wind of modern education be contained? The *sutradhar* of the Foundation said: By the act of faith in tradition, by the act of will in transforming, surely it can.

From the flowering of all these thoughts the fourth question arose. What would be the seed and the soil of the new school? The *sutradhar* explained: Education is the art of intelligent living; it involves the training of the mind and the awakening of the heart. Deeper humanity emerges when the mind is trained and the heart is awakened. It is wise to remember that the seed of universal education lies in the mother's womb and blossoms in her lap; it is cultivated in the soil of nature and culture, kinship and community; and it brings with it the ideal of beauty, goodness, and happiness.

From the inner idea of the playwright, the fifth question arose. What can pervade the mind of the spectators? The *sutradhar* replied: The emotional and other states, as created by human nature and in accordance with the ways of the world.

The Opening Scene

The *sutradhar*'s movement in time began with chaos and despair. Within a few days of taking over the Shikshalaya opposition came from the two inmates. A group of Bengali women staged an angry protest. Not so surprisingly, the woman who used to clean the premises led the demonstration. In the midst of utter confusion the new management began its mission — 'the education of heart'. However, much of what happened in the beginning ultimately served as a shock absorber.

All the teachers of the old school were Bengali. They were retained but were told that no amount of dry discipline would do the children much good. In those days Muslim and Harijan children were not allowed to use the toilet, which was meant exclusively for the inmates and teachers. This was changed forthwith. Muslim girls above eight years used to attend the school in *burkah* (veils). They were asked to give up the veils within the school premises. The response was encouraging. After a few months no one came with the *burkah*. The attendant of the school was an old lady who had been given a low position. The headmistress of the school, a *suchibai* (purity-pollution maniac), treated her as an untouchable even though she was a Bengali Brahman. When the school attendant was invited to inaugurate the Independence Day celebration everybody present was shocked. This kind of radical change subsequently caused a silent stir inside the Shikshalaya.

A few months later there was a theft in the premises. What the thieves left behind was more dreadful than the loss of property: the footprints of a child followed by larger ones. After some time, another theft occurred. The drunkards, the gamblers and the goondas began to play their role of villain. At this stage a suggestion came from one of the members of the Bose Foundation that reformative measures should be initiated first to better the people of the locality whose children were to be educated in the new school. The *sutradhar* responded that the results of the reform would be seen only after two decades, when the children of the school were grown.

The next year, the old teachers left the school *en masse* without notice. By that time the number of students had sharply declined, and people got the impression that the school was going to be closed down. A wealthy trader in silk fabric came out with the suggestion of opening a *madarassah*-type school for which a suitable endowment and some two hundred children would be made available to the Foundation. The benevolent proposer was told that the Foundation would be happy to teach Urdu, Arabic or Persian and would also impart religious education in Islam, *provided* that the children received instructions in the arts, elementary science, mathematics, Hindi and English. This was unacceptable to him. Later he offered the Foundation a plot of land (double the present space) some 4 to 5 km away, a new building, and five lakh rupees as compensation for the campus. He was told that his pious resolution had no value because the objective of the Foundation School was to make education self-organizing and self-supporting and to develop aesthetic sensibility and cross-cultural understanding among both Hindu and Muslim children. He was assured that as soon as the school was firmly directed towards its goal and the neighbouring community became strongly inclined to take up the responsibility of running it without government aid, the Foundation would withdraw and would repeat the experiment in another area of the city.

The Foundation was prepared to brave the worst.

A Kashmiri *pandit* was staying at the Foundation as a guest scholar. He was working on Kalhana's *Rajatarangini.* A Persian teacher used to call on him frequently. Their relationship appeared to be normal. But a stage came when the teacher began terrorizing his student because he refused to be converted to Islam and married to a widow. The seventy-year-old *pandit* had no courage to face death in the present crisis. He left the Foundation in fright.

The air of the city was poisoned by communal bitterness and rancour following the court's verdict on the temple-mosque issue in Ayodhya, politicized by sections of Hindus and Muslims. All educational institutions were ordered to close down. When the Foundation School reopened after a few days, the Muslim children came with black strips on their arms. They did not know why their parents made them wear this colour of sorrow. For these innocents it was a mere fun and fashion. The school was sunk in shame and sorrow. Not a word was uttered on this subject. The bell rang for the prayer meeting. The usual duration of silence was prolonged. In silence the *sutradhar* and teachers prayed for purging all hearts of communal hatred and ill will. Within no time the Hindu and Muslim children rolled into one and the black strips disappeared completely unnoticed, as the darkness of night is dispelled by the sun.

The Melodies

From chaos and despair emerged new thoughts and activities like melodies in a concert. At the Foundation School *swaraj*, *swadeshi* and *sarvodaya* — the three arts of life and education — began sounding together in harmony.

Swaraj in education means "the self-rule in the management and administration of an educational institution". Most people today feel utterly unable to run a school without a government grant-in-aid. Alternatively they turn the school into a commercial enterprise. If the desire to establish *swaraj* is genuine the problem will have to be faced boldly, not replacing government rule by the rule of merchants. The teachers who came to the Bose Foundation School merely to earn a living without the spirit of service found it trying. Many of them left within a few months; many insisted upon applying for a government grant-in-aid; many wanted expansion of classes or grades with a view to increasing income through pupil's fees; others suggested conducting special tutorial classes on extra payment. Nothing of this kind was conceded. The failure of their attempts caused many experienced teachers to leave the Foundation School. Happily, in course of time it was found that some youths from middle-class families developed a fascination for the School, despite the fact that they were barely paid a living wage. The Foundation is proud of having teachers like Rama Lahiri and Rajesh Iyer, who have served it lovingly for more than six years. Freedom, friendship and trust are the strongest bonding factors that give the workers at the Foundation School amazing courage to face collectively the music of life. Looking back, it can be said that the Foundation today demonstrates the ancient idea of teachers teaching for the love of it and receiving the barest maintenance.

The Foundation School has no corpus fund, no annual grant, not even irregular grants from any source. It is run on school fees and royalties on the books published by and for the Foundation. It is not a profit-making venture; it has learnt to live in holy poverty. The fact that it has managed in this manner for over sixteen years strengthens the Foundation's conviction that primary or elementary schools must necessarily be self-organizing and self-supporting. Community and religious trusts could provide a corpus fund for the life of the school. If education means the awakening of consciousness, if the children have the freedom to grow into fine human beings, and if cultural diversities are to be respected, then all primary schools must be managed by those whose children are to be educated. No centralized nation-wide system should operate at this level of education.

Swaraj in education means *swadeshi* in spirit. In the domain of education *swadeshi* implies that it is a virtue to remain firmly grounded in the perennial wisdom tradition of one's own culture. Education is essentially a cultural process of making human life efficient and complete. Hence indigenous institutions are the best to impart education at the early stage in life. As Gandhi said, "Swadeshi is not a cult of hatred". *Swadeshi* spirit in education means the spirit of expansion, not by driving out ideas but by absorbing new ideas without suffering cultural identity crises.

As a research centre, the Foundation receives distinguished scholars — specially anthropologists, sociologists, linguists, philosophers, historians of religion — both Indian and foreign. Their presence has contributed immeasurably to the awakening of the children. For several years the students of the Wisconsin College Year in India Program were affiliated with the Foundation. Their first step on entering into the Foundation was to interact with the children of the School. They practised Hindi on the children and in return widened the perceptions of those who had never seen the world beyond the physical space of Kashi. In this interchange, friendships developed. The children invited them to their homes. Most of the students of the Wisconsin Program studied various aspects of Kashi's cultural traditions. One of them studied the children's traditional games, another worked on the riddles known to the children. Joyce Hubert wrote in 1983 a dissertation based on an anthropological study of the Bose Foundation School. Some of them taught English to the children of classes IV and V. For the last six years the Foundation School has been a training-cum-study centre of the University of Karlstad (Sweden) for its teachers in education. Inger Wiklund, the co-ordinator of this scheme, has been a Visiting Teacher at the Foundation

Self-learning

High concentration

Learning together

The little monitor

Children are nurtured in a liberal and lively atmosphere

Education through the Arts

Festivals are the occasions of performance – making things of beauty, doing higher things.

Acting old

Dancing
at the spring festival

Becoming the
goddess of learning

Teachers truly great

Youths of the middle class families from the neighbourhood demonstrate the ancient idea of teachers teachng for the love of it and receiving the barest maintenance.

Give children the attection of a mother or a sister. Children address them by kinship terms, elder sister or brother.

School since 1991 (see Appendix-1). Indian scholars, Gandhian social workers, and religious personalities are frequent visitors to the School. This constant exposure affords a highly refined sense of perception that brings great confidence to the children and ultimately purity in cultural order. It is this way of education that enables the children of the Foundation School to live with the *swadeshi* spirit in the extraordinary world of ideas.

Swaraj in education aims at *sarvodaya* — education for all. A votary of true education cannot subscribe to the utilitarian formula of the greatest good of the greatest number. He must strive for the greatest good of all. Every child must be educated enough to observe morality and to attain mastery over its mind and its passions.

The Foundation School brings transformation from within tradition; it maintains a high standard of teaching and tries its very best to serve the desire of parents, who demand instruction analogous to that of the modern mission schools or the commercialized public schools. But it does try in its own way to suit the capacity of poor parents. Life at the school is simple. The neighbourhood is poor. The school stands by what is implied in the Gandhian phrases 'Sarvodaya' and 'Antyodaya' — Unto This Last.

The Joys

With the declared objectives of *swaraj*, *swadeshi* and *sarvodaya* in and through primary education, without reference to government grants, the Foundation School is organized on an experimental basis. An attempt is made, with limited resources in men and money, to evolve an operative model rather than develop a conceptual framework for self-organizing education. The school has founded a sacred tradition of working — the right way of doing right things. It is a new *gurukula* which, in contradistinction to the old *gurukula*, is a non-residential institution. But it has all the advantage of the old indigenous system that combines the home and the school in one. The school is located in the heart of a mixed community of Muslim silk weavers and Hindu castes following a number of modern and traditional occupations. Its building matches other residential buildings in the neighbourhood, providing at a casual glance the picture of an extended family. As it makes itself a part of the neighbourhood, the children do not have the problem of adapting to a different environment. The neighbourhood is economically poor but culturally rich. So is the school: it cannot impart free education to all. Moreover, it also holds the view that payment of school fees makes guardians realize that education is as essential as food; it gives them confidence. Free municipal schools hold little attraction. The children of the Foundation School pay their fees on time. The poor have learnt to live with pride and dignity.

True education is inseparable from religion. The Foundation School considers morality the first principle of religion. It follows the approach of Kabir and Gandhi — the gospel of love, the belief in the oneness not merely of all human life but of all that lives. It emphasizes the oneness of God and the brotherhood of man. It shares Gandhi's dream.

Prayer at the Foundation School is undoubtedly the highest expression and application of religion. It is combined with meditation and silence. The School teaches respect for parents, respect for tradition, and love of the motherland. The children are told that the mother is the first and foremost *guru* and that their motherland is the daughter of Mother Earth. They are taught to respect all faiths and to come forward to imbibe knowledge of the religions of other peoples. They are made aware that the religion of humanity, i.e. 'universal religion', does not stand in opposition to 'practised religions'.

The religion of humanity is not social ethics but the art of life. Art is the expression of aesthetic experience. Works of art are, therefore, not works made for profit but essentially works of goodness, beauty and harmony. Through the practice of art, a child not only achieves creative power but also enjoys learning. Religion, art, and education are essentially inseparable in their doing and feeling. The Foundation School opens in the morning for three to five hours but the children keep coming in the afternoons to play and in the evenings for singing and dancing. Festivals follow Nature: the cycle of seasons having both physical and spiritual sensitiveness. Festivals are the occasions of performance — making things of beauty, doing higher things, holding exhibitions, and feeling fun. They are also occasions for singing and hearing music, dancing and acting — *not* really for competition but primarily aimed at the delight of the mind. Seasonal celebrations at the Foundation School are not purely religious; they are concerned with the illumination of beauty and the lightning of intelligence.

There is a difference between the intelligence of man and the intelligence of the machine, or artificial intelligence. The machine does not have a sense of beauty, a sense of goodness, a sense of love. It is a designer of things, not an artist of the imagination. The Foundation does not consider education a tool to design commercial things. Education is an intelligent imagination for which beauty, goodness, and love are the foremost needs. At the primary level of education what is required is to impart knowledge of the nature of things and to develop the faculty of intelligent imagination by which the child can act. The Foundation School endeavours to do this by introducing thirty minutes of prayer-cum-meditation, regular yogic exercise, and the art of poetry recitation and story-telling without the use of classroom texts.

For the children of classes IV and V, it is made obligatory to write a notebook, called *Jnanamanjari* (the flowering of knowledge), where their favourite stories, songs, poems, puzzles, descriptions of festivals and other cultural events, etc., are recorded and illustrated with pictures. Experience has shown that they mirror the minds of the children and help the seed of their thought grow. A content analysis of 117 *Jnanamanjari* notebooks written over fifteen years reveals that these children, regardless of their being Hindu or Muslim, share in common the knowledge of Kashi that gives a certain universality — the Third Culture of this city — far beyond the frontiers of their respective religious orthodoxies.

What makes a teacher truly great is imagination combined with knowledge, understanding and affection. The teachers at the Foundation School are of three categories: regular teachers, visiting teachers, and monitors. Under the last category, senior pupils work as assistants to teachers. Past students are also involved as full-time teachers. Experience has shown that formal training of primary school teachers is of little value. The words 'trained teacher' mean 'made by deliberate skill' rather than the product of instinctive operations or impersonal force of nature as manifest in the mothers of all animals. The best teachers are those who give children the affection of a mother or a sister, who play with them and tell them stories and riddles. The teachers are addressed in kinship terms.

At the primary level, printed books create a degradation of the mind. The Foundation School does not make a fetish of literacy. The children of this sacred city have an awareness of the history and culture of India through shrines, temples, mosques, fairs, festivals, *lilas* and rituals. They are also aware of the different kinds of people of the world visiting the city as pilgrims and tourists. For the children of the Foundation School the world is not a book of maps, and the cosmos is not a concept but a living reality that presents itself in Kashi, the microcosm. Their minds are prepared to know the real thing and to see the beauty in things around them, so as to reach nearest to the highest human value.

The Echo

Sixteen years ago the Foundation began sowing the 'seeds' of ideas in the hope that they would bear fruit. The wise man cultivates the field of knowledge without knowing that he knows. So is the experiment in education. In either case there is a great joy of working. The joy of working at the Foundation School lies in seeing and experiencing that the soil is excellent. It is thrilling that Muslim girls fettered by convention to profit by new educational systems (*Figs.* 1-3) now stand first in all the seven classes (Table 1) and first in the School (Table 2); their customs forbade them in the strongest manner never to go outside the house without the veil, but they come to the school unveiled and excel in public performances in the arts (Table 3).

There is an *echo* of change in the environment (Table 4). The children of this school have learnt to restrain communal passion. Their guardians, especially mothers, have realized the value of school education. Those who had rejected the Foundation's vision of cross-cultural education now see a culture of higher things evolving. Those who once used to throw stones at the Foundation School are now the most frequent spectators of the children's performances. In their direct experience of working, the *sutradhar* and teachers find the flowering of their service, the feeling of heart, the hope of resilience, the harmony of the rhythms of life and knowledge, and a noble culture of peace evolving.

But what is achieved in these years is of no consequence — it is only the beginning of a tradition in cross-cultural education. It will take generations to purify and stabilize the new system tried out at the Bose Foundation School.

Table 1(a): First in Class 1981-89

	1981		*1982*		*1983*		*1984*		*1985*		*1986*		*1987*		*1988*		*1989*	
Class	H	M	H	M	H	M	H	M	H	M	H	M	H	M	H	M	H	M
A	-	B	-	B	G	-	G	-	-	B	B	-	B	-	-	B	B	-
B	-	G	-	B	-	B	B	-	G	-	B	-	G	-	G	-	-	B
I	-	B	-	B	B	-	-	G	B	-	B	-	-	B	-	B	B	-
II	B	-	-	B	B	-	B	-	-	G	G	-	-	G	-	B	G	-
III	B	-	B	-	B	-	B	-	B	-	-	B	G	-	G	-	B	-
IV	G	-	B	-	B	-	B	-	B	-	G	-	-	G	B	-	-	B
V	-	B	G	-	B	-	B	-	-	G	B	-	G	-	-	G	-	G

H = Hindu, M = Muslim, B = Boy, G = Girl

Table 1(b): First in Class 1990-95

	1990		*1991*		*1992*		*1993*		*1994*		*1995*	
Class	H	M	H	M	H	M	H	M	H	M	H	M
A	G	-	-	G	B	-	B	-	-	B	-	B
B	-	G	B	-	-	G	B	-	-	G	-	G

(*Contd.*)

I	-	G	B	-	B	-	-	G	-	B	-	G
II	B	-	-	B	-	G	-	G	-	G	-	B
III	G	-	B	-	-	B	-	G	-	G	-	G
IV	B	-	G	-	B	-	-	B	-	G	-	G
V	G	-	G	-	G	-	B	-	-	B	-	G

H = Hindu, M = Muslim, B = Boy, G = Girl

Table 2: First in School

	1985	*1986*	*1987*	*1988*	*1989*	*1990*	*1991*	*1992*	*1993*	*1994*	*1995*
H	B	G	G	-	B	G	B	B	-	-	-
M	-	-	-	B	-	-	-	-	G	G	G

H = Hindu, M = Muslim, B = Boy, G = Girl

Table 3: First in the Arts

Year		*Drawing*	*Dance*	*Acting*	*Singing*
1990	H	-	-	B	B
	M	B	G	-	-
1991	H	B	-	B	B
	M	-	G	-	-
1992	H	B	-	B	B
	M	-	G	-	-
1993	H	-	-	B	B
	M	G	G	-	-
1994	H	-	-	B	B
	M	G	G	-	-
1995	H	-	-	B	-
	M	G	G	-	G

H = Hindu, M = Muslim, B = Boy, G = Girl

Table 4: The Echo of Change in the Environment

A. Beneficiary of the Foundation School 1980-95

Religious Groups	*Families*	*Students*		*Total*
		Boys	*Girls*	
Hindu	236	596	493	1,089
Muslim	194	362	342	704
Total	430	958	835	1,793

B. Response of the Guardians

A sample survey of the guardians' awareness of school education was conducted in 1995 by Ms Rama Lahire of the Bose Foundation School.

Hindu (40%) and Muslim (60%) respondents in the age-group 28-60 answered a 16-point questionnaire. Of them 70% were women, 42% illiterate, 33% having Madrassah-type schooling, and only 25% with up to Class VIII education. Most (72%) of them belonged to the low middle class families and the rest lived in abject poverty; 84% were involved in craft and 16% had taken up small jobs.

On the question of school dress, English education, engaging a tutor, religious education (both in home and in school), girls' education, small family, and combining education with work, 100 per cent respondents gave a positive answer. Muslim informants wanted that their children may learn Sanskrit and Urdu, but English they must. Among the silk weavers there was no abhorrence for traditional occupation; they wanted their children to follow the ancestral trade.

On the preferability of the Foundation School, their considerations were: proximity, cleanliness, safe for girls, well-behaved teachers, good atmosphere, generosity for poor guardians, teaching of arts and other good things, moral education, paying attention to pupils as evident in their progress — developed sense of responsibility, intelligence, sharing, enthusiasm, and discipline.

STUDENTS FIRST IN CLASS

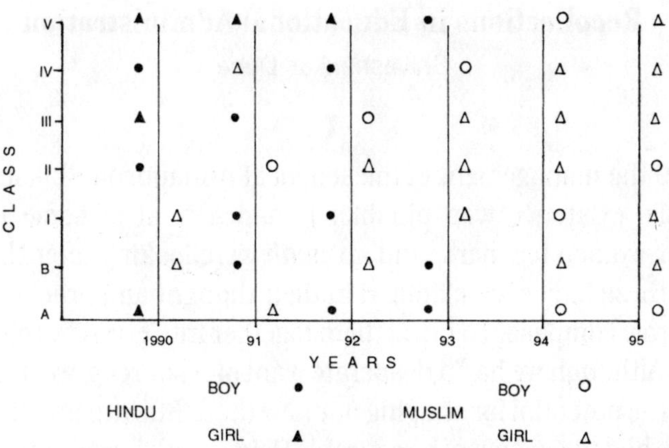

Fig. 1

MUSLIM STUDENTS FIRST IN CLASS

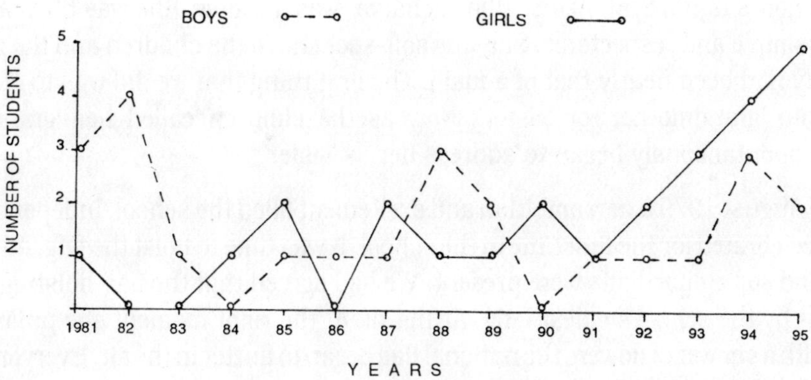

Fig. 2

FIRST IN SCHOOL 1990-95

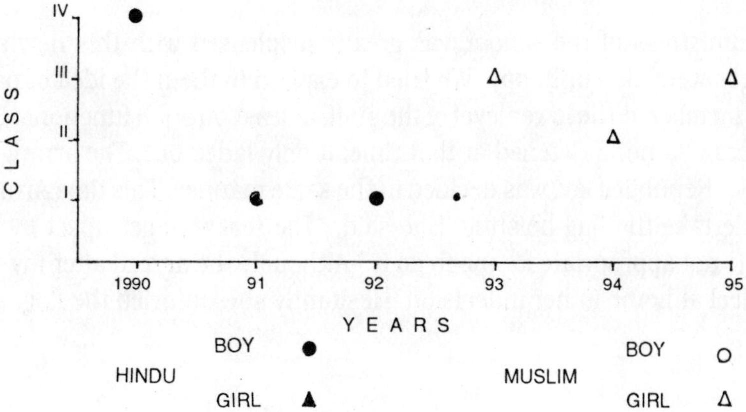

Fig. 3

Recollections in Educational Administration

Shivashankar Dube

I

When in July 1979 the management of the school at Annapurna Shikshalaya was handed to me, the school's existence was pitiable. It had a total of some fifty children. The headmistress, four women teachers, and an *ayah* were looking after the school. Working within four walls, these ladies were limited in their thought and practice and were used to living within a narrow compass. To bring them together into our new thinking and purpose was a difficult job. Although we had a desperate want of resources, we had dedication in our thoughts. We had the potential for shaping our new thoughts into practice. More important was the fact that we had the benefit of honourable Baidyanath Saraswati's guidance. Despite our resourcelessness we were, therefore, full of a new energy and excitement. The school was named Sarojini Vidyakendra after the founder of the Annapurna Shikshalaya.

The school's *ayah*, Smt. Aruna Bhattacharya, was a widow. She was 55-year old. Her living was simple and respectable. She was soft-spoken. To the children and the guardians her identity had been nearly that of a maid. The first thing that we did was to change this identity into 'grandmother' or 'sister'. Now as the children called her 'grandma', the guardians spontaneously began to address her as 'sister'.

On 15 August 1979 a new ambition and excitement filled the school. Independence Day was to be celebrated for the first time in the school. It was time to hoist the flag. All teachers, children and some guardians were present. We had agreed that the flag-hoisting would be carried out by the school's oldest lady, Aruna-di. At the right moment she pulled the flag strings. With a shower of flowers the national flag began to flutter in the air. Everyone saluted the flag. The practice of flag-hoisting by a government officer or a well-known person, according to the school tradition, thus ended and a new practice as well as a new tradition began.

The headmistress of the school was greatly displeased with this new arrangement. Other teachers were also unhappy. We tried to explain to them the idea of paying respect to the oldest member at the lower level of the staff, at least on such functions. But the anger of the teachers was not quietened at that time; it only faded out. The arrangement for 26 January 1980, Republic Day, was decided in the same manner. This time Aruna-di was not ready to undertake the flag-hoisting. She said, "The teachers get upset by it. I also feel unhappy. It is not appropriate for me to do it." Although she agreed after my words to her, she did not feel at home in her indecision. Hesitantly she unfurled the flag.

II

The first spring festival was arranged for January 1980. A three-day programme was organised for art exhibition, dance, drama and music. All guardians, children and residents of the neighbourhood were invited. The school volunteers were occupied with keeping order. Police assistance had not been asked for. Spectators made quite a crowd on the first day. The crowd began to press for want of space. At the same time some uncivil persons began to come in. Some were intoxicated with drink. Fear and apprehension began to grow in the spectators. The situation went beyond the volunteers' control. Quarrels started. In their apprehension the spectators began moving away. The programme had to be postponed midway. The situation became quiet after some pleading.

It was necessary to take police help in desperation on the second day of the programme. Admission cards were introduced to limit the crowds. Those who could not enter began throwing stones into the *pandal* during the programme. The programme started under police vigilance. Now a new problem appeared. A Muslim girl who had been trained in Bharatanatyam had given a performance on the previous day and the spectators had acclaimed the dance. On the second day the same dance could not take place because that girl did not turn up. I went to her home to find out the reason for her absence. The guardian spoke about her aching stomach, and therefore she was instructed not to take part in the programme. For us it was a great disappointment because she has prepared the dance with great industry and expertise.

After a few days it became known that the girl was really not indisposed. Her father had forced her to stop taking part in the programme. A few Muslim neighbours had seen the girl's dance on the first day. They told her father that singing and dancing went against their tradition. The privacy of their honoured heritage had been injured. The result was that the father prevented the girl from performing. Not only that, he also objected to her schooling. Her two younger brothers in infant classes were transferred to another school.

Earlier the girl's father had proved a keen governor for our school. In his unselfish service he had also taken up the responsibility for teaching Urdu to the children at the school. This episode, however, knocked us in many areas. Three children of our school went to another school, and their father stopped teaching Urdu. It was widely said in the neighbourhood that the school provided little education and wasted time in teaching singing and dancing.

Lesson learnt. For the next year's spring festival we included children in the programme only after obtaining the consent of their guardians. Despite this some children involved in the rehearsal for the programme went out of town at the time when they were required to perform. The guardians did not take it seriously. In this way we had to confront in the beginning many such obstacles in relating the children to cross-cultural performances.

III

Our objective is not to provide bookish knowledge in a formal manner but to make efforts to bring out the creative talents hidden in the children. Holding subsidiary classes in music, dance, drama and art as well as competitions in them are the efforts in this direction. As a result of this initial enterprise the children presented on the stage the impressions of their inner feelings with a kind of ease and simplicity from which we came to know that the natural genius for art is latent in every child.

My experience is that not only are the feelings and thoughts of the children at our school different from those of pupils in public and convent schools, there is also a relatively greater degree of naturalness in them. Perhaps this is because these children are not the offsprings of competing households.

From sixteen years of experience we have come to understand that in all forms of teaching the role of music, dance and drama is important. Most of our children come from the household environments of artists and craftsmen. We study them first before we try to teach. They remain strangers to haphazard and quick learning. We try to relate the children particularly to prayer, meditation, music and dance, acting and sports. On the basis of our experience we can say that in all these involvements an atmosphere is created where the children are 'disciplined' on their own, and where there is discipline the idea of controlling is rooted out.

Although we confront even now many obstacles to relate the children to artistic manifestations, the parental opposition that used to appear during the early years does not exist today. One of the most important reasons for this is that our old boys and girls, who are now learning in upper classes, encourage their younger siblings to participate in the cultural programmes. If there is any opposition from parents it is sorted out by itself. Not only that, if their brothers and sisters are not selected for any important competition, the old pupils are disappointed.

IV

The school's second year began. It was July 1980. There was much pressure on the school in this month. The headmistress was ill. She was on leave from 11 July. I was facing a lot of difficulty in dealing with the children's enrolment and TC. On 13 July another mistress was absent without notice. The school was somehow running with three teachers. I also went to teach the children in an unengaged classroom. On the second day I went to see the headmistress at home. I advised her to come to school only after she was well. She was worried by the news about the school.

On 15 July no mistress had arrived at the school by 7 o'clock. Aruna-di also had not come. The children had come and had become noisy. I made a child ring the bell for the prayer. All went to sit in the prayer hall. I was worried. Aruna-di had just come. She put four envelopes on the table and said, 'Read them, *bhaisaheb*'. I opened the envelopes and read the letters. They were from four mistresses about a collective resignation.

I went to the prayer hall in poise. I completed the prayer and meditation. The children were serious. Of the teachers I was the only one present before them. I could understand their seriousness. Breaking their questioning silence I said, "Children, accidentally your sisters (mistresses) are on leave for urgent business. So, no teaching will be possible today. You let go home and study your lessons there". The children went away.

This put me into great difficulty. I informed my friends Baidyanath Saraswati and Ramlakhan Maurya about events after they had come to the school. We had never imagined that the mistresses could bring themselves to submit their resignation in unison. We went to the headmistress. We informed her about the incident. She wondered why it happened. She made no comment but let us go with the assurance of her presence at the school the next day.

We decided that teaching must not be suspended even for a day. The young pupils in our own households were selected for this.

Next day the school opened at the right time. Some enterprising young boys and girls had been collected from our own households to teach the children. Among them Kumud Maurya, Rajesh Shrivastav, Sudha Dubey and Divya Maurya should be named especially. I was waiting for the headmistress but she did not come and instead her resignation arrived. I felt depressed by her conduct.

The entire situation became clear. The design to put the school into trouble had been arranged beforehand and in this the role of the headmistress was the most important. She had nine years' experience in the management of this school. I had learnt quite a lot about managing the school from her within the brief span of a year. I also had the intention to learn much more from her in future. Her departure from the school in this way with all her fellow mistresses was very unfortunate. In running the school I always moved with her before me. I had tried to keep the mistresses in a firm self-confidence to explain or understand our new efforts and purposes. Although certainly there were occasionally differences of opinion, we respected one another at the practical level. We had taken up the school as an experimental institution. For us it was a challenge in the area of primary education. The old mistresses condemned this challenge and confronted us with a resignation in unison with the claim that the school would not operate without them. We bowed to them. To call them back was recommended. We had, however, accepted everyone's resignations.

Today there is no place for a headmistress at our school. The mistress who has, however, acquired the image of dutifulness and dedicated service among her colleagues, has come to be seen as the senior-most mistress. In everyone's heart grew a love and attraction towards her. The mistresses addressed her as *Barididi* (older sister).

V

Initially the school was running on its very limited resources. With the change of management we were forced to confront serious financial problems. The mistresses were taking sixty rupees every month as salary. We set the salary to a scale and made it eighty rupees every month. The increase was met with donations. The following year we tried to solve this problem by increasing the number of pupils at the school. According to an earlier arrangement the school was getting some regular financial assistance from the Sharda Sangha. The Sangha stopped paying our institution. In terms of our constitution we could not take assistance from the government. Our foremost challenge was this terrible situation of financial deficiency.

We did not appoint teachers with prior advertisement but invited talented, enterprising and dedicated young men and women. The intention of love, respect and good feelings was prominent in this invitation. The teachers came to us not to work but to serve. The tradition is continuing even now. Is it possible to imagine such dedicated teachers living in respectability when they are paid only three hundred rupees every month? It is an inconvenient question for us. At the moment the total number of children is 125. Our fees for a month are forty rupees. About twenty-five children are exempt from fees. In comparison with the fees at other schools our monthly fees are lower by a half. Although we feel strongly that we should not take fees from such poor children, we cannot help it. The fees are our only support.

The money that is required for festivals, cultural programmes and to prepare the children for their participation in competitions is entirely separate. Our teachers and children meet this expenditure from the collection of donations, although this flow of income has become slender.

It has never been possible to keep a teacher on a regular honorarium for dance, drama, music, painting and yoga. We are left with no extra money for them. We are able to give them only their cost of travel. It is only the honesty and dedication of such teachers that have enabled our children to give us memorable performances. We also receive the cooperation of some social organisations and workers influenced by our performance. The arrangement for the toilets, furniture, fans, water, etc., is the contribution of such organisations.

VI

We have two problems, internal and external. The want of dedicated teachers and finance is the principal internal problem. Today the teacher wants to teach as an employee. There is no feeling of service and dedication to education at all in this. We have to retrain the teachers and mistresses we select to our requirements. When they become capable in their service to the school a few years after the retraining, they become impatient to leave the school for various reasons. If one leaves for permanent employment, another leaves after marriage. We do not receive any benefit for our labour and time. We cannot pay any honorarium to the teachers of dance and drama. With only the cost of travel they offer their services voluntarily, but under pressure of responsibilities have to leave the school.

The role of guardians and immediate neighbours appears before us as the external problem. We take the children's guardians into three types. The first are the guardians from among the illiterate, the poor and labourers. The number of their children is the highest in the school. Most of them are devoured by such bad pastimes as gambling, alcohol and lotteries. They are oblivious to their children's education. They think that once the children are put into the school, the school has the full responsibility for them. The second type of guardians, who are somewhat literate, continue to try to get their children into the school at the secondary level. As soon as they have succeeded in this they remove their children from us. They argue that in this way they are free from the hassles of the children's enrolment to secondary school. The third type, who have some education and whose financial status is also a little firm, get their children enrolled only so that they come to school and go back home. When their children have learnt something in a year or two, they send them to the English-medium convent schools. We made a programme to run every month a committee for such guardians. However, we did not receive their cooperation in this. Only eight or ten guardians could come to the meeting. The committee closed. The non-cooperation of the guardians is an obstacle for us.

The children's environment and neighbours are also relevant as problems. Most children up to class V come from illiterate and poor households, chiefly of 'untouchables' (tanners) and weavers. Such a household is enclosed with a dense population and filthy environment. The children from such households play or loiter in dirty alleys after school hours. They are friends with the local loafing children who do not attend school. Their guardians do not understand the influence of the environment. They only know that the school is responsible when a child has failed in his study. They think the school is careless about their children's education.

VII

Here are a few actual experiences of shaping the children.

CASE 1

NK is an old pupil, a poor boy from an 'untouchable' caste. He used to sell rags and discarded *saris* with his mother after school hours. He passed class V in 1981. After studying elsewhere until class VIII he devoted himself fully to his own business. Despite this he maintains his connection with the school. Whenever there is a festival at the school, he comes uninvited and helps.

Once Aruna-di had become ill. He used to go to see her everyday, and took fruits, bread, biscuits, etc., for her. No one knew about it. When I went to see Aruna-di I came to know from her about NK's sense of service. One day the doctor asked for Aruna-di's phlegm to be tested. It was raining heavily. NK came to me soaking. Aruna-di had sent me a phial of phlegm with him. I began to wonder how it could be sent for testing in the torrential rain. NK gauged my worry. Despite the pouring rain he was ready to go to the hospital. Although he was told not to go out in this weather, he did not agree. He said, "Master saheb, Aruna-di has served the school so sincerely. Can't we do even this little for her?" He went to the hospital in the downpour.

It was the school's spring festival. Money was needed for choreography. I was worried. NK came to know about this problem from a teacher. After a while he came and put five hundred rupees before me. I could not understand at once. Laughing, he said, "Master saheb, do the festival with this money. I'll be getting some more money by evening". I was only too happy with this modest and dedicated seventeen-year-old pupil.

CASE 2

GK and HA were brought for enrolment into the school. Because of the difference between their ages and education it was learnt on inquiry that they had already studied at several schools in the neighbourhood. They used to play truant. Instead of going to school they would wander and play around. They had been expelled from several schools for this reason.

The guardian was told to come to the school everyday to report on both and they were enrolled in class I and Infants A on this agreement.

The children's mistresses were briefed fully on them. They were told to keep an eye on them as well as treat them with affection. At the same time, they were told to avoid too much pressure on learning and too much discipline.

After talking to both children with affection an attempt to win their confidence was made. An occasion to talk to them everyday was arranged. They had a great liking for sports. So they were given full facilities for sports at the recess. To organise sports they were made captains for football after a few days. They were given full responsibility for taking out, putting back and letting the children play with the ball.

Gradually both children began to come to the school regularly. They would come to the school premises half an hour before the school session. Taking out the football they would play a little and join the prayer in time. Whenever they found time from study they would play their game. When the school was closed for holidays, they would come frequently for the game. No hindrance was offered by the school to their love of sports.

After nearly a month one day the two children were seen at the school before time. However, they were absent at the time for the roll call in the classroom. A teacher was sent out at once to look for them. The two were discovered in a park in the company of a number of unknown children. They were brought to the school. With the assurance that they would not be punished for truancy and the guardian would not be informed, the reasons for their truancy were asked for. They said that with the temptation to take them to the pictures some of their old companions had encouraged them to leave out school for the day. They had collected the savings from their tiffin. The old companions who were the school's discards had threatened them because they did not take them to the pictures.

The children revealed the entire story truthfully. They were advised to leave the bad children's company and were also told that if in future they were threatened or harassed they should come to the school straight away and report the matter.

I made clear to them the results of the evil company of bad children and rewarded them for telling the truth, because they had the opportunity to lie. GK was made the prefect for his class. With this making an astonishing change appeared in his conduct. Both children gave up the unknown children's company. The regularity of their attendance remained satisfactory. They also improved their performance in study. They continued with interest in keeping themselves in the front with the school programme, such as cleaning the classroom, putting the children's shoes in a series, placing the children in line at prayer time and ringing the bell.

Both children remained at the school for three years. Their guardian was quite satisfied with them. He said that the same children also used to steal money at home. Now they had changed radically. They helped with the housework. They had given up the wayward children's company and remained engaged with regularity in their study.

At the moment both children are carrying out important responsibilities in their family trade (weaving).

CASE 3

KM (10 years old, Class IV) used to study in his village. He could not take an interest in study. His everyday routine was fighting with children at the school, tearing up exercise and reading books and running away from the school whenever he liked. His family was harassed by his waywardness. He was chastised more and more severely at school but his impertinence kept increasing. Finally he was expelled.

KM's grandfather was a teacher at Banaras. He brought the boy to Banaras to educate him. He wished KM to get into our school. He gave a full report on KM. He believed that KM would change here.

Next day he came to me with his grandson. I asked KM a few questions. We talked. He had learnt nothing from his village school. Although he was not suitable for enrolment into the Infants, I took him into Infant-A though he was much older than the rest of the class. It was a complicated problem to teach him with little children. I told his grandfather, "If you take sufficient interest in KM, it will be possible for him to change".

KM began coming to the school. His impertinence started as well as the children's complaints. I used to make him sit next to me. He would listen to me attentively. Frequently he would come to school late. The teacher used to warn him. I would talk to the teachers about KM's problem and explain that he did know that he was coming to the school late everyday. We should keep him away from other children and keep watching. He has changed partially. He is coming to the school everyday. His grandfather had frequently complained that KM made a lot of mischief at home. Everyone was harassed. Only I could bring him round.

Finally KM's householders got so harassed by his mischief that they sent him back to the village. It was discovered only later that KM had not attended the school for several days.

After a year KM's grandfather again came to me with him. I warned him that my efforts would bring no fruits until they took his problem seriously. His grandfather assured me that he was now leaving KM in my care and I would get full support from him.

I understood that this time KM's indiscipline was more complicated. I took him into a particular routine in my own control. I tied his responsibility to my own class. I left to KM the responsibility for cleaning the room everyday, the arrangement for drinking water and the circulation of notices. He continued doing all this with interest. Sometimes I would leave the office to get away for a while. After coming back I used to see KM sitting in his own place like a sentry reading or writing. I used to ask him to bring me anything he could write at home on his own. He would bring his writing and show it to me. I found in him the ability to act with a sense of honour. After a few days, at the children's meeting I praised KM's work and placed before the children's conference the proposal that KM could take the responsibility for ringing the bell for the prayer. The children passed the proposal happily. KM began to ring the bell every morning at seven. His grandfather complained to me that KM went to the school frequently without breakfast and that I should correct him. I made it clear to his grandfather that he ought to understand KM's responsibility and send him after giving him his breakfast at the right time. He might sacrifice his breakfast for want of time.

The boy was not absent even for a day in ringing the prayer bell. As a result of this a surprising change came over him. He began studying his class lessons on his own. He was selected as the prefect of the class.

Four years have gone by and KM is a pupil with this year's class IV. He is now given entirely to his study. For attendance and regularity he is an ideal pupil at the school. He is also responsible for the equipment for the children's games — carrom board, ball, ludo, etc. In all programmes at the school KM's role is always significant. For sports and acting KM is the school's prize pupil. The children love him.

All would like KM to leave the school only after passing class V.

CASE 4

DJ (14-years-old, Infants-A). His father drives an autorickshaw. He has a stepmother. It is a middle-class family. DJ was accustomed to steal other children's tiffin to eat. At the tiffin break some children get occupied with sports. Their satchels remain in the classroom. After playing for some time they take tiffin. Some children frequently complained that someone stole their tiffin and ate it. For several days the mistresses did not feel concerned about it. However, when the complaints began coming regularly, they became watchful. Some children also complained about theft of erasers, pencils, etc. People started looking for the thief. Eventually DJ was found one day stealing tiffin. The mistresses admonished him. However, he could not give up his practice. He became well-known to the children as the 'tiffin thief'. When no admonition made him get rid of the habit, a complaint was forwarded to his guardian. He was given a beating at home but he could not correct himself. One day his harassed class mistress forcefully brought DJ to me with a crowd of children. All requested me in one voice to send him out of the school.

I quietened everyone and directed them to their classrooms. I had DJ sit with me.

Through discussion with him I tried to understand his thoughts. I found out from him that his stepmother frequently took money out of his father's pocket. Often the husband and the wife quarrelled violently. DJ knew that his mother used to take money out of his father's pocket in his absence without his consent and that she considered this act right. So DJ had understood that it was not immoral to take others' belongings without their consent. He did not understand it was theft. It seemed to me that DJ had no thought for stealing, but only in his curiosity he opened another's tiffin and seeing food before him, he ate it. For the tiffin he brought with him used to have good things to eat and they were always in sufficient quantity. The same was true for the erasers, pencils, books and exercise books in his satchel. No material was ever scarce.

I did not reprimand DJ for stealing tiffin. I asked him to take his tiffin with me everyday. I said that he would leave his tiffin with me from the next day and at the tiffin hour we two would have tiffin together.

He informed his mother about it and next day brought for tiffin *puri* and vegetables made to his instructions. When we were taking tiffin sitting together at the tiffin break, other children began looking at us with great interest. The children became a crowd. It was strange that DJ was taking tiffin today with the big 'sir'. On the other hand, in his pride from a comparison between him and other children in taking his tiffin with me DJ was smiling, looking towards the children. 'See, your friend also wants to taste your tiffin. Serve him as well'. DJ jumped up at once and put *puri* and vegetables into the mouths of two or three children. The children moved away from us sheepishly.

The routine of our taking tiffin together continued for weeks. Meanwhile, I used to praise his tiffin and say, "Your tiffin is always the best of all childrens". I used to check his satchel and say, "You have got everything, books, answerbooks, erasers and pencils, and they are the best of all". He had given me a beautiful pencil as a gift. Gradually I began encouraging DJ to give a share of his own tiffin to the children who used to call him 'tiffin thief'. Sharing tiffin began to make DJ feel happy. With his fingers DJ would put tiffin into the mouths of those who hesitated.

After a few days I began to send DJ to his fellow pupils to take his tiffin and observe the responses it produced in him. I saw DJ had become a tiffin giver, giving and taking tiffin with other children with a happy expression. He began attending particularly to the children who could not bring tiffin from home. He would give them a share of his tiffin and request other children also to share their tiffin. DJ never again stole tiffin. Later he turned into a cooperative and industrious student devoted to his study. He passed class V with the Vidyakendra's first class. At the moment he is a student in the class VIII at a different school.

His greatest peculiarity is weaving on the looms after school hours at his leisure.

CASE 5

AJ (12-years-old, Class III) used to read at the school in Nownihal. He became friendly with bad children. He used to call mistresses names. He would come out of home for the school but wander around in school time with strange children. He would reach home when school ended. For this he got beatings at school and at home. He was unable to correct himself. A vexed family stopped his study. AJ enrolled in class III at our school. The guardian had already told me everything about him. One day AJ took five rupees out of a girl pupil's satchel. The pupil complained to the class mistress. Everyone was questioned. However, nothing was discovered.

I asked AJ in private. In tears he admitted to stealing the money. Sobbing very loudly he said, 'Master saheb, don't tell anyone at home, please. I'll get beaten up'. I quietened him. I congratulated him for telling the truth. I convinced him that I would tell no one about this. I asked him, "Why did you leave your old school?" He said that he got too much beating there. "When someone's belongings were lost, only I was punished. I could not give myself to the study there".

One day AJ found fifty paisa on the school premises. He brought it to me and said, "Master saheb, someone has dropped this money. Please find out and give it to him". I spoke about his honesty at the children's meeting. I rewarded him with an answerbook and a pencil.

AJ is now a class V pupil. He has surpassed everyone in physical labour. Only he is responsible for the school garden. His guardian says that he is mad about plants and flowers. He has also got flower pots at home. He spends his pocket money only on them. He does not show any bad habit any more.

Recollections in Workmanship

Ram Lakhan Maurya

When the Bose Foundation took up the management of the old school at the Annapurna Shikshalaya, we had intended that this primary school would become a centre for multiple talents. We wished the seeds of knowledge, the sense of nationality and the best human qualities to grow in our children. Our chief aim was to shape the form of initial education for the children, born in distressing poverty, with an initiation into a feeling for beauty in them, encouraging the understanding of their inner culture and its purposes, developing in them new images of initial education as well as initiating them and their mistresses into willing and lifelong service to humanity.

The challenges were big but our ability was limited. When the Bose Foundation had started the school, it was not getting any financial assistance such as grants from the government and other sources. The school was running on its own limited efforts and little income. After more than sixteen years even now we do not ask for government grants or financial assistance. People feel surprised that this school has been continuing for almost two decades despite its financial difficulties.

Around the school live poor weavers, oppressed labour connected with the Banarasi *sari* enterprise, 'untouchables', and a financially invalid and suppressed collection of people from the scheduled and backward castes. Most children at our school come from such families. Most cannot study beyond completing primary school. Their parents have neither the liking nor the means for their children's education. Now an incident from the last month: with a two-year-old child in her arms and holding the fingers of an eight-year-old girl, a veiled Muslim lady from a weaver family came to get the girl enrolled into the school. It was now the month of September and children were enrolled in July. She said that the girl's father and grandparents did not care to have her educated. The parents inquired, as they never studied nor did their parents, what this girl would do after her education. They went on ignoring it every year with such statements. The girl is already eight years old. If she was educated she would have some abilities and wherever she would go she would make her famous, otherwise she would lose her reputation. Out of exasperation she had therefore come out of home. She heard that in view of Gandhi's centennial, enrolment was made this year without fees in this Foundation School and so she had come with her.

Our greatest challenge was to develop a liking for education in the children and guardians from an environment confronting financial problems, to grow a mutual understanding in children from different faiths and languages as well as cooperation and accord, and to diffuse knowledge in the atmosphere of their traditions.

So far in the school's unrestricted management the willing cooperation of some of our dedicated voluntary workers has been a great contribution. In our school greater attention is given to the teaching of dance, drama and music. In the celebration of *rituparva*, the festival of the season, we have built a tradition. At the spring festival cultural programmes with dancing, singing and drama performances run for two days. Our children take part in the competitions in drama and the other arts. They have become well-known everywhere for their excellence in cultural programmes. At many competitions at the city level our children have been awarded the first place in competing with children from large and well-known convent and English-medium schools. Trophies have been taken.

We have never introduced salaried teaching for the children's art education at our school. Residents of Banaras established in the arts of dance and music come to our school to train the children without salary.

For general education also we have the cooperation of many great teachers. Among the teachers invited to our school are academics from universities and colleges as well as learned scholars from India and abroad. They have been teaching the children without salary. As an invited teacher Inger Wiklund of Karlstad University in Sweden has been teaching the children English for the last four years.

Wherever former pupils of Sarojini Vidyakendra may be studying, they maintain an intimate relationship with their old school. Whenever a celebration or assembly is arranged at the school, they come uninvited and give us their assistance. It is our former pupils who take the responsibility for tasks from stage-building and choreography to the welcome of the guests.

On the last full moon the former pupils of the school gave a new expression to their own dedication and interest in service to the school. In July 1995 some pupils together opened the former pupils' association (see Appendix-2). Now their number has increased to over fifty. On this occasion many programmes of service to the public were announced. They also announced free classes in the evenings for the educationally weak and backward children of the Sarojini Vidyakendra and from the very next evening classes began operating. In this our age such unselfish service from children is rarely witnessed.

The kind of experiment with which this school is involved is difficult and time-taking. But we have good feelings about the work. I wish to place on record some of my most memorable experiences.

The school's spring festival was close. The children were occupied with the preparation of their own programmes. A girl called Parvin Banno was playing the role of Sona, the dancing girl's daughter in the historical drama *Deepdan*. She was always engaged in practising the dance. I saw her rehearsal always and encouraged her to work hard. She had made her role the object of her reputation and went on practising for hours.

On the first day of the festival *Deepdan* was staged. Parvin had filled her role with life. The spectators were thrilled with her dancing. When the curtain came down after the dance, the entire makeshift auditorium rang with the clapping of hands. Parvin's happiness had no limit. Moved by her own success, she came running breathless to me in full make-up among the spectators and asked, "Master saheb, how was it?" I was unable to say a word and kissed her forehead in blessing.

Now the story of a day. I was taking the lesson in moral instruction for the children of class V. I made them understand the greatness of obeying one's elders. At the end of my talk I said that children should obey their parents. I asked a pupil if he followed his parents' instructions. The pupil stood up and said, "Sir, my father returns home drunk every night, quarrels with my mother and beats us too. Sometimes he sends me to get his drink on credit. Should I obey his instructions?" I could not do anything with him at that time except express my sympathy. During the midday break I said at the day's meeting with the mistresses that it was not enough to explain a moral to children with good words. We ought to take into consideration our personal conduct as well, as they learnt through the imitation of their elders.

A child in the nursery was in the habit of eating other children's lunch without their knowledge. When some children would get busy in their games on the grounds at the midday break, this child would take out their lunch and eat it. One day he was caught. The children took him to the teacher. He was told that one must not take another's article without consent. Stealing was bad. When a child close by heard the teacher, he said, "Sister, in his childhood even Lord Krishna used to steal curds, but people worship him". The teacher was unable to answer.

Sarojini Vidyakendra, in Gauriganj, Bara Gambhir Singh ward, is a sensitive area. After the destruction of the Babri masjid the condition of this area became even more delicate. However, the environment has no particular influence on our school on the borders of this area. People connected with both communities respect this school as a multi-denominational centre.

At the time of Saraswati *Puja*, the worship of the goddess of learning, children from every community at our school install and worship the image and make offerings. On the occasion of *Holi* an assembly is arranged and so also for *Id.* At *Id* we also visit the children's homes, offer our greetings and consume the *sewai* dish.

Once at the time of *Id* the atmosphere was a little tense. None of the teachers could visit the children's homes. At the school the next day the *ayah* told the teachers, "A few children had come with *sewai.* They were very sad when they could not get any of you. They have left *sewai* for all of you". I felt a happy satisfaction in me with the sweetness of the children"s feelings.

An incident from the last *Holi*. The school was closed after the completion of the *Holi* celebrations. It was evening, the dust of coloured powder was floating in the air above the streets and the atmosphere was filled with excitement. At such a time it was not safe for women to go out of home. However, two Muslim girls of our school went out with coloured powder and arrived, crossing the lanes and the streets, at the home of a schoolmistress about four furlongs away. Seeing them so late in such an atmosphere the schoolmistress was very worried. She asked, "Why did you come out at this time?" A girl said, "Sister, you were not at the school for the *Holi* celebration. That is why we have come to see you".

The schoolmistress wondered what their parents would have thought if anything had happened to such innocent girls. Surely they had come out without their parents' consent. The mistress gave them some snacks to eat and sent them home with her brother.

An established, well-known and rich Muslim family engaged in the *sari* trade lives in the neighbourhood of our school. We came to know that the head of the household had just returned from his pilgrimage. The school decided to invite this religious person to meet the children. His presence was requested on a special day. The children collected in the Gandhi room at the school. Some of our friends and colleagues were also present. A famous social worker friend and follower of Gandhi and Vinoba welcomed him and delivered a long thoughtful talk in his honour. The children garlanded him. Those who came with Haji Saheb also took photographs. Haji Saheb was filled with emotion. It was the first occasion in this area when such a welcome and assembly had been arranged at a school in honour of a person on his return from pilgrimage.

Haji Saheb narrated his recollections of the pilgrimage. At the end he promised the school a donation of twenty-five rupees every month. We thanked him profusely for his generosity.

After a few months an image of the goddess of learning was installed at the school on the occasion of Saraswati *Puja*, as in other years. The children worshipped for two days and received the offerings. During those days Aruna-di was the *ayah* at the school. She was an old lady without artfulness and was loving and attentive to all with a very clean heart. She thought, "The school received a donation of twenty-five rupees from Haji Saheb, who is also a member of the Vidyakendra community. As everyone received the offerings, he should also get some". She went to Haji Saheb's home with a plate of offerings. She came across Haji Saheb accidentally. She said, "I've brought offerings from Saraswatiji. Please take them".

Haji Saheb was filled with anger. With a scowl he ordered her to go out of the premises. Aruna-di never understood an affront. She was unable to understand the cause for Haji Saheb's displeasure. Little did she know that in Haji Saheb's religion, offerings to the deities of other religions were considered sacrilegious! She came back crying.

From that day Haji Saheb stopped his grant of twenty-five rupees to the school.

Of the children's sweetness, love, kindness and fellowship as well as human qualities there are many sweet and sour experiences which cannot be narrated within the limits of paper.

APPENDIX-1

The Bose Foundation School: Seen Through Western Eyes

Sarojini Vidyakendra

Inger Wiklund

I have been working as a visiting teacher at the Sarojini Vidyakendra (SVK) The Bose Foundation School, since 1991, teaching English, participating in all kinds of cultural and extra-curricular activities, observing the working of the school or simply delighting in its unique atmosphere. Here I would like to present in short some of the ways in which I have found that SVK differs from other schools, Indian as well as those of my own country, Sweden; ways in which this humble and materially poor school could serve as a model to the richer schools of the developed world.

Small is beautiful: Education on a small scale. Although it would be much wiser from an economic or commercial point of view for a private school like SVK to admit large numbers of students, SVK insists on keeping classes small, with usually not more than twenty students in a class, there being only one class or section in every grade. Small and few classes contribute to an atmosphere of informality and affectionate intimacy in the classrooms. All teachers know all students, and every individual student knows and interacts with all other students of the school.

School as a family: The feeling one gets at SVK is not that of a school, of a formal educational institution, but that of a large family. The students are all sisters and brothers to one another, the older children naturally guiding and helping the younger ones. Teachers are referred to as older sisters, *didis*. SVK reaches out to the homes of its students and establishes bonds and relationships that usually last long after the children of a particular household have left school. In this way problems of 'weak' or 'troublesome' students can be traced to their roots in their homes, and encouragement is given to and pressure exerted on families to let their children go on to higher education after leaving SVK.

Teaching as service: The teachers at SVK are encouraged to see teaching as a service to humanity and not as a way of making a living. They do not receive a regular salary but a very small honorarium. As a result, the staff of SVK is always made up of honest, sincere and dedicated individuals whose selfless work seems to have made a definite impression on the students. On their own initiative, two young former students of the school are now teaching and giving free tuition at SVK.

Cultural integration: Hindu and Muslim, high and low caste children, all study together at SVK.

The school works actively to eradicate any feelings in the children of communal, religious or racial divisions and to instil in them instead an appreciation of basic human values. Religious festivals of all communities are celebrated collectively at the school.

Emphasis on culture and the arts: All schools include the teaching of the fine arts in their curricula but SVK, following Tagore, places an unusually large stress on the teaching of music, art, dance, drama, etc., as a means of making learning attractive to the students. Several months may be dedicated to the preparation of a cultural programme, awakening in these children from culturally poor homes a love of the fine arts.

The Children of Assi

The Transference of Religious Traditions and Communal Inclusion in Banaras

Marc J. Katz

Since 1986 I have observed over one hundred hours of classroom in Banaras schools. On only two occasions did I witness teachers dramatically engaged in traditional-style story-telling without the use of classroom texts. Both of these teachers were unusually active in their teaching methods and naturally outgoing in personality. One of the teachers was at the Sanjay School and taught the second and third grades where she used this story-telling method with vigour, but only rarely. The other teacher was well-known for her story-telling ability and often engaged the smaller classes of the N.K. Bose School in 'activity' interacting with her stories. She even encouraged her students to interrupt her performance and make observations and put relevant questions to her. Occasionally her story-line would turn into a song. She gesticulated with her arms. She interchanged serious drama with humour. Her story-telling methodology was highly reminiscent of the traditional *katha* story-telling performances that are an everyday occurrence in the temples and on the *ghat*s of Banaras. She claimed that she was encouraged by her colleagues. And although her story-telling method was almost identical to the well-known traditional performance of *katha*, she was considered unusual. I was told that she was 'experimental' and definitely in line with the intentions of her employer.

The N.K. Bose School is an educational institution for both Hindu and Muslim children of poverty. The school was started by the anthropologist, Professor Baidyanath Saraswati, who lists the school's primary objectives as "the development of aesthetic taste and cross-cultural understanding and evolving new perspectives in primary education". Dr Saraswati refers to his school as a 'lab school' where it is possible to experiment with varying teaching methods.

A Socio-Anthropological Micro-Study of Sarojini Vidyakendra

An Experimental School in Varanasi

Joyce Hubert

Sarojini Vidyakendra is an experimental school administered by the Nirmal Kumar Bose Memorial Foundation, a foundation for socio-anthropological research.

The Foundation operates on the principle that instruction, via the collection and assessment of data, can lead to constructive action. To achieve this, an outreach programme to the children's homes is aimed at amassing vital statistics about the students' families and homes; and at gleaning the religious and social attitudes that prevail there. The administration is also keenly interested in testing the children to determine the extent of their supra-rational and rational training.

To foster unity in the children's lives the administration has devised a school day that is four to five hours long. They encourage the student to utilize the remainder of the day assisting their families in the family business or with household duties. The interviews revealed that this is indeed the case: all of the children participate regularly in the maintenance of the household.

Dr B.N. Saraswati, General Secretary of the Foundation, has stated that social advance is evident in the comparison of traditional and modern educational techniques. Both techniques are employed at the school. Classes in traditional dance, drama, and music reinforce traditional religious values. Many of the songs that are taught to the children are Hindu devotional songs; the dances are for the most part narrative in style — enactment of the *Ramayana* or other stories about Hindu gods and goddesses. Other songs and dances have regional and folk sources. Some of the texts are socially oriented: the lyrics extol the virtues of hard work and adequate nutrition, or relate the official ideology of India, i.e., Muslims and Hindus can and want to coexist peacefully. The modern educational devices are in the form of class methodology. The students are permitted a fair amount of freedom, which is a Tagorian concept. They are also taught Western science, English, and mathematics. In this way they receive instruction relevant to the socio-political exigencies of the contemporary world.

The educational ideology behind Sarojini Vidyakendra lies between that of Tagore and that of Gandhi. Between the ideologies of these two men, one highly sensitive to the realities of physical suffering and the other sensitive to disembodied suffering, lies the ideology behind Sarojini Vidyakendra. By interweaving data, empirical evidence, educated analysis, and philosophic ideas, the Foundation aims at the evolution of a learning situation which is at once pragmatic and poetic.

Books are distributed to the students during class time and are collected at the end of each class. Many of the books are torn. They are stored in a cabinet in a dark, dank cubicle that functions as the library, teachers' sitting room, and kitchen. The few musical instruments that the school owns (wooden percussion instruments, a harmonium, a few tambourines) are also stored in this room.

In this cubicle, a delightfully colourful elderly woman, employed by the school, prepares *chai*, snacks, and sometimes meals for the administration, teachers, and guests. She does this squatting upon the floor over a kerosene stove situated next to a water tap. During 'tea time', about midday during the student's recess, the teachers sit upon a low wooden bench in this room and drink the *chai* prepared for them.

No meals are provided to the children. However, they can purchase glucose biscuits from the kitchen employee for a few paise. Otherwise, the children must wait to eat until they return home at the close of the school day.

The school day varies with the seasons. During the hot season the school day begins at 7.30 and ends at noon. During the cold season, December until March, when the school is too cold before the sun has warmed the building's walls, classes run from 11 until 3. Classes meet six days per week. Each day is divided into seven half-hour periods, and a recess of one half-hour. The latter was somewhat surprising to learn, as ignorance of the school structure led me to believe that the class duration depended upon the attention and interest of the students. The latter is partially true, and if the students are particularly interested they will remain in the class after the gong has been struck indicating the period's end.

The children's families pay monthly fees of ten rupees. For this small fee the children derive some of the benefits of a private school, i.e., small classes and individual attention. However, for the families of the children, ten rupees are not so easily attainable.

Although the physical aspects of the school are dismal, the spirit is warm and bright. In fact, upon first exposure to the classes and general atmosphere, the school seemed to be comfortably chaotic. The children do freely move around during classes, especially during music and performance classes, but generally their activity does not seem to be detrimental to the learning process.

Corporal punishment, i.e., physical discipline by striking a student, is illegal in many of the United States, and if not illegal in other states, a substantial number of people find the practice to be abhorrent. When two of the teachers were informed of the position of corporal punishment in the U.S., they exhibited surprise and curiosity. One teacher indicated that she thought that corporal punishment was necessary in India: 'The students are so wild'. The second teacher did not verbalize her opinion, but observation of her limited use of the stick reveals her position.

Whether the teachers enjoy teaching or not, whether it is morally correct or not to strike a child, is almost secondary to the fact that the children enjoy themselves in the school. They frequently linger at the close of the school day, until they are asked to leave. Without hesitation they met us at the school on Sundays, to take us to their homes. They are quite comfortable coming and going from the school at any time during the day or week. Never have I observed either in government schools or private schools in the U.S. such informality in the school situation as I have at this school.

Saturday is seminar day. On this day classes focus on performances. Students perform dances, songs, or recite poetry for each other and their teachers. At the first seminar that I attended, the student performed individually before the other students, administrators and guests. The performances seem to be by rote, as they performed with their eyes averted from the audience. However, it has become apparent over time that the students' confidence increases as they perform repeatedly before their classmates and teachers. The rote memory comes alive with the interaction between the memory and its life before others.

I have observed very few academic classes. Those that I have observed, with the exception of one class, have been English and math classes for the younger children. The children were seated on the floor as they transcribed by note method from the textbook into their copy-books. When their work was complete they presented it to the teacher, who was seated by the door, for correction. The only additional academic class that I did observe was a math class for classes IV and V. The children were namby-pamby throughout the room: some were standing, others were seated. The teacher wrote mathematical problems on the blackboard. It was an exercise in algebraic thought. The children were expected to solve the problems without the aid of pencil and paper; they would shout their guesses to the teacher. Much excitement ensued when the teacher explained the mathematical principle.

- An interesting exercise that we did in the workshop was feeling hands. Here some people were wrapped in a blanket with only their hands showing. The students had to feel their hands and narrate their experiences. One of these was a man whose only job was to water the lawns of NID with a huge hose-pipe, another was a young girl school student, and there was one student from NID itself. Students had interesting experiences to tell here, particularly of the man who watered their lawns, whom they had seen day after day on their way to their classes but never spoken to. Suddenly they began to understand him.
- We had a vertical mural where students could put up anything they liked on the theme of hands. In the centre of this mural was a big hand drawn by one of the students.
- There was a horizontal panel on which students could come up with something on the chosen theme and pin it up. In fact registration to the workshop was the hand-print of the student on a sheet of paper pinned up on this panel.
- Then there was a sculpture box. This was basically the wooden crate in which fruit is sold. Each one of the students was given one of these and asked to make something inside or around it. I must narrate one very interesting effort by a girl. She put up a screen in front of the box. One could not see what was beyond it. And she said one had to feel with the hands the objects inside. We could feel inside clay, water, some relief work that she had done and other materials that she had used, giving it an extra dimension that does not meet the eye.
- Students were asked to create something on the theme through pottery, painting, collage, photography, and so on in their studio hours.
- Finally, on the same theme of hands, the students also did plays. The plays took them two full days to make. They used mime, visuals, story, music, and so on.

The entire workshop was for seven days. We worked only on one element — the hands. Sometimes individually and sometimes together but everything grew from within them. In a sense I did not 'teach' them anything.

Learning Through Instinct

Probably one of the greatest sources of learning is play, and *prakriti* has great lessons to offer which can be unearthed, discovered with this play. A child simply has to play. Without playing he either cannot remain a child or cannot learn anything. In fact everything can be

The children were solemn and reserved during the ceremony. They observed a ten minute meditative interlude in absolute silence . . . quite surprising to see a group of four-to eleven-year-old children silent for such a prolonged period.

APPENDIX- 2

Founder Members of the Pupils Association of the Bose Foundation School, Varanasi, July 1995

Sl. No.	*Name*	*Address Bhelupura Ward*	*Schooling years*
1.	Sri Vijaya Kumar	B3/198 Shivala Ghat	Five
2.	Sri Manoj Kumar	B7/29-5B, Devariyabira	Seven
3.	Sri Om Prakash Pal	B8/26 Bara Gambir Singh	Four
4.	Ms Parveeen Bano	B9/37 Gauriganj	Five
5.	Ms Suraiya Bano	B9/37 Gauriganj	Four
6.	Ms Meenu Kumari	B9/51 Gauriganj	Seven
7.	Ms Padma Kumari	B9/51 Gauriganj	Five
8.	Ms Laki Kumari	B9/51 Gauriganj	Seven
9.	Sri Nasir Khan Lodi	B9/61 Gauriganj	Five
10.	Ms Tasneema Rizavi	B9/68 Gauriganj	Four
11.	Sri Ranjeet Chatterjee	B11/1-A Gauriganj	One
12.	Sri Satyajeet Chatterjee	B11/1-A Gauriganj	One
13.	Ms Amita Chatterjee	B11/1-A Gauriganj	One
14.	Sri Amjad Khan	B11/19 Gauriganj	Seven
15.	Sri Manish Sharma	B12/8-B Gauriganj	Five
16.	Sri Bablu Visvakarma	B12/19-A Gauriganj	Five
17.	Sri Shamashad Khan	B12/23-A Gauriganj	Four
18.	Sri Bablu Jayaswal	B12/36-A Gauriganj	Seven

(*Contd.*)

Sl. No.	*Name*	*Address Bhelupura Ward*	*Schooling years*
19.	Ms Baby Naz	B12/43 Gauriganj	Seven
20.	Ms Soni Rani	B12/62 Gauriganj	Six
21.	Ms Munna Rani	B12/62 Gauriganj	Six
22.	Ms Farha Khanam	B12/62 Gauriganj	Four
23.	Sri Deepak Jayaswal	B12/67 Gauriganj	Seven
24.	Ms Aparna	B12/80 Gauriganj	Five
25.	Sri Ramesh Kumar Chaudhary	B15/69 Faridapura	Seven
26.	Sri Mahesh Kumar Chaudhary	B15/69 Faridapura	Seven
27.	Sri Suresh Kumar	B15/69 Faridapura	Seven
28.	Sri Anil Kumar Verma	B20/1-9 Bhelupura	Seven

References

Gandhi, M.K., 1938, *Hindi Swaraj or Indian Home Rule*. Ahmedabad: Navajivan Publishing House.

———, 1947, *India of My Dreams*. Ahmedabad: Navajivan Publishing House.

Gill, Eric, 1983, *A Holy Tradition of Working*. Suffolk: Golgonooza Press.

National Policy on Education 1986 (with modifications undertaken in 1992). New Delhi: Ministry of Human Resource Development, Department of Education.

Newsletter, 1978, 1, 1. Varanasi: N.K. Bose Memorial Foundation.

Nurullah, Syed and J.P. Naik, 1951, *A Student's History of Education in India*. Bombay: Macmillan & Co. Ltd.

Saraswati, Baidyanath, 1975, *Kashi: Myth and Reality of a Classical Cultural Tradition*. Simla: Indian Institute of Advanced Study.

Sinha, Surajit, 1986, *Nirmal Kumar Bose: Scholar Wanderer*. New Delhi: National Book Trust.

Vaudeville, Charlotte, 1974, *Kabir*. Oxford: Oxford University Press.

5

Ghadatar

An Enquiry into the Invisible Order

Haku Shah

Learning through the Unconscious Mind

In Gujarat, particularly in Saurashtra and Kutch, a girl embroiders everything for her to be born after she marries and goes to her husband's house. She embroiders a cap for her child. Here she thinks of the basil plant that she worships everyday. She dreams of horses, elephants, birds and figures — women with water pots or women with flowers — all the visuals that she lives with, she remembers and wants to remain with her child. Look at the physical process in which the woman is involved. It is like learning through the unconscious mind, imparting to oneself one's own thinking through skill, creativity and, in that sense, love.

The process of dreaming of the child to come starts much before the child is actually conceived. Once conceived the process becomes more focused, clear. When a child is born it has already accumulated all the above through the mother and the world around it. Thus lessons for a child begin even before its birth. That is why it is essential for parents to take cognizance of it. We assume that the age of learning begins at five to six years, but if we really think of children's education then we must take into consideration the time when the child is in the mother's womb. A very tender seed takes shape and gracefully grows.

If parents would behave themselves whilst the child is growing before it is born and after, it is a well-known fact that the child would instinctively obey the law of truth and the law of love.[1]

Once the child is in this world all its senses become active and thus the great door of the universe opens. At this stage the second chapter of learning begins. Here two questions may arise:

1. How do we know that the child is learning?

2. What is it that the child is learning?

Do we remember the baby gazing at the mother or the mother gazing at the child while she feeds it? In a very short while the child gets to know of the mother: 'this is mine'. It may have learnt many more things than just the fact that she is its mother. When the mother is feeding the child we notice the interaction in the silence, the talk and the *ami drishti*, or the fondness with which they look at each other. All these are lessons for each other — not only for the child but for the mother as well.

So school has already begun for the little one — from being fed by the mother, to her stories and jokes, the sounds, the *ami drishti*, the dreams, the conscious and the subconscious mind are very much at work. After a while, when the mother dotingly says 'you cunning', which actually means 'you darling', this abstraction is communicated and understood by both of them.

The Process of Inquiry

When the mother sings a lullaby she sings of plants, flowers, birds, and people in the family, and even sleep is called out to — '*nindar rani* (Queen of Sleep), come and put my child to sleep'. Here she also reproduces the sounds of different birds and animals and invites them to play with the child in its dream. Sometimes when the child becomes older, it asks for a particular song of which it has become very fond. This lullaby then remains with the child for a long time and plays a role in defining its character.

A teacher is a mother. She who cannot take the place of a mother can never become a teacher. A child should not feel that it is receiving education. The child whose mother's eyes follow it everywhere is receiving education all the 24 hours.[2]

The self-learning process begins early. The child learns to say '*ma*', mother, as the first word. Then follow words like '*pa*' for father, '*na*' for no, '*ha*' for yes, and so on. Learning from sign language also begins here. For instance, you point a finger and say '*jo*' or 'see'. All these and many others build up the learning process of the child. Sign and word language become more and more interesting. When the child utters the first word it becomes a great source of pleasure for the parents. So much so that they even start imitating the child and the child in return enjoys repeating the same word often. When the child becomes a little older, the parents start talking to it in its own tone. Sometimes the mother speaks to the child as if it were a grown-up person.

The process of enquiry begins. The learning which takes place through the senses now begins with questioning. 'If grandmother is dead, where has she gone?' 'Where have I come from?' The child wants to inspect everything, break things and see how they are made, touch what they are asked not to touch. Two-year-old Jai was asked, 'What is your name?' He replied immediately, 'My name is Jai but my mother calls me *atankvadi* (terrorist)'. Sometimes he takes out a cardboard box, empties it and beats it like a drum. His mother comes up to him and says 'Jai, will you stop bothering me with all that noise and go to the neighbours and play with your friends?' On another occasion he did the same thing again with the box, but as his mother entered the room he said 'Don't worry, mother, I will not bother you with the noise and will go next door and play with my friends.' If one analyses the two instances the knowledge and the concept of faith that the child acquires from these informal situations at such a young age is an indication of the procedures in imparting education. To the same boy, then, we teach the entire gamut of do's and don'ts:

Do not tear paper, or you will destroy it

Do not play in the dust or you will get dirty

Do not play in water, you will become wet

Do not play in the sun or you will get tanned

Do not climb the tree or you will fall

All don'ts will only hamper the positive process of creative enquiry in the child. The attitude of parents and teachers cultivates in children a sense of fear and cowardliness, of untruthfulness.

Where there is untruth or fear there is no love. Love is the key for learning. Without it learning or education is impossible. Having been blessed with children I discovered that the law of love could be best understood and learnt from little children. Were it not for us, their poor ignorant parents, our children would be perfectly innocent. I believe implicitly that the child is not born mischievous. Whilst the child is growing, before it is born and after, it is a well-known fact that the child would instinctively obey the law of truth and the law of love. We put aside the law of love and truth and in the name of education, behaviour, good habits, and manners we cultivate fear, hate untruthfulness, etc., in children. Where there is fear there can be no intelligence. So intelligence comes into being with the understanding of your own self and you can understand yourself only in relation to the world of people, things and ideas.[3]

All the time we order the child: you should sit and read . . . you should obey your elders . . . you should learn from your elders . . . do not argue with your elders . . . sit and study for your exams . . . . There is nothing unknown to the child and at the same time it would like to know about everything. Sometimes to get the mother's attention to a question the child even holds the mother's face while being asked to sit quiet 'when elders are talking'.

Often we want to see our child reading and writing quickly. We are in a hurry to get that into the child even if it is a burden on him. Unless we remove the notion that the child will gain knowledge only through reading, the process would become difficult. It can easily be conceived that he who has no knowledge of the alphabet throughout his life may become learned.[4]

A child can draw a human figure — his mother or father — very early in life. But the great art teachers would think otherwise. 'You should first learn to draw the leaf', they would say. The child can express itself through available material marvellous things, whether it is the sand on sea beach or a desert or any such thing. Unfortunately many parents say that 'he has no inborn talent to be an artist'. Dr Coomaraswamy once said that an artist is not a special kind of human being but man is a special kind of an artist, or else he is something less than a human being. So don't we want our children to be human beings! Then we should give them a chance to paint, draw, sculpt, sing, dance or create whatever they want.

Once an illiterate singer, Ganesh, came to me. He was about thirty years of age. In a sense he was like a child. He had never held a pen or a pencil. I asked him to draw. 'I don't know how to hold a pencil', he said. 'Try: you will be able to do it'. 'Why do you want me to do this?' 'For me! I would love to see how you draw'. He agreed and started drawing. 'Make a tiger'. 'I do not know how'. 'If not a tiger then make an elephant'. He drew an elephant and a man riding on it. 'Who is this man?' 'Myself'. He had actually made himself!

Then he started drawing more and more. Once when a friend of mine and I asked him to draw the story of his life, to our surprise, while depicting shops in Ahmedabad, in his drawing he started making letters and numbers which did not make any sense — like the name of shops. Along with that he drew a sample of what the shop was selling . . . a dress or a radio or some such thing. So we knew what the shops were selling.

Then I asked his wife to draw. And surprisingly, within a few months' time she started drawing 70-feet paintings of goddesses! Both husband and wife loved to paint. Even their children started painting. To our great surprise, they made the most unusual themes like themselves; what they sang in their songs; sounds of birds, *shahnai*, etc.; dark nights, planets. Now here we may ask: Where is the real alphabet in this? Is it learning? Is this knowledge?

The Self

The question that arises in my mind is: Why can't one, whether literate or illiterate, create one's own lessons? Their own poems and their own plays? When we give children prescribed books to read, they would surely learn what the author has to say, but what happens to the child themselves? When we were children we were given textbooks to read as if they were the end of the world of knowledge and there was nothing beyond. And so the self is completely lost.

A teacher teaches as though he or she teaches were superhuman. They do not realise that there is so much that they have to learn from the students as well. The teacher cannot see himself and does not let the students see within — that is the greatest tragedy. The other being, of course, that no matter how little informed one may be of one's subject, one can always pass examinations by hook or by crook and end up attaining a degree, losing out in the process the sheer joy of learning the truth, the freedom and the expression of the self. How can one learn like this?

When I tell students to be true to themselves when they create anything, to do what they think is best, they say, 'What do you want, Sir? We'll do what you want, Sir'. They are conditioned to obey and act accordingly.

One can say that somewhere in this process the self gets mortgaged. The identity of the self is lost. Although it remains somewhere, it is taken over by teachers, advisors, preachers, parents, and so on.

In some of the temple cloth paintings done by the Vaghari folk painters of Gujarat, their gods are placed in the centre and minor deities, musicians, stories, ancestors, environment and others occupy the space around. I showed my students slides of these paintings. They had to design anything on this concept based on someone they loved, with other things around. The students came out with marvellous works and it was great to see whom or what they loved. One created 'mother', others a child, a television, chewing gum, etc. They actually got to think of whom they loved.

If you could read all the books in the world it would not give you intelligence. Intelligence is something very subtle; it has no anchorage. It comes into being only when you understand the total processes of the mind — not the mind according to some philosopher or teacher but your own mind. Your mind is the result of all humanity and when you understand it you do not have to study a single book, because the mind contains the knowledge of the past.

You can learn from books but that does not take you very far. A book can give you only what the author has to tell. But the learning that comes through self- knowledge has no limit because to learn through self-knowledge is to know how to listen, how to observe, and therefore you learn from everything — from music, from what people say, from the way they say it, from anger, greed, ambition. The leaf that is blown by the wind, the murmur of the waters on the banks of a river, the flight of a bird high in the air, from the poor man as he walks by with a heavy load, the people who think they know everything about life, you are learning from them all. Therefore there is no teacher and you are not a follower.[5]

Self-knowledge comes through the process of self-inquiry.

The Invisible Order

Right from birth the human being brings with him an invisible order. This then gets expression through visuals, sounds, action and movements. This order exists in each one of us. We see this order when Ganesh and his wife, Teju, draw. When you ask a painter to draw sounds he might be taken aback. But when Ganesh and Teju drew, it was without any hesitation, through sheer intuition.

Once Ganesh decided to narrate the story of a potter in one of his drawings. He was inspired by a story often told in Gujarat. In this story a potter lives with his wife in a village. In their house they also have some cats. One day one of these cats gives birth to kittens and puts them in the potter's half-baked pots for safety's sake. The potter's wife gets to know of this and is pleased about it. The next day the wife leaves for her village, a short distance away. After she has gone, oblivious to the kittens being inside them, the potter puts his half-baked pots into the kiln all ready for baking. The wife sees the smoke rising from the kiln and comes running back home in horror to save the kittens. The potter and his wife fear it might be too late already. They wait in great fear for the kiln to cool down. He pulls the pots out of the kiln. Much to their astonishment and relief, they find the kittens alive and cuddled up in the pots.

Now Ganesh in his drawing made the burning kiln and the pots and so on. Then he came running to me saying, 'Oh! But I forgot to remove the kittens from the pots, what should I do now?' I told him to make nearby a drawing where he could show that the kittens had been removed from the pots as in the story, and found to my amusement that he happily accepted my solution.

Sometimes one wonders that for someone who has never been to school or had any kind of formal education, one who did not 'learn' in our sense of the word how to draw an animal, an elephant, a camel, and so on, how can he draw? Which of his faculties helps him draw here? Is it that he sees well, is it that he perceives well, is it that he can observe well, or does it come from intuition?

In many ways he is like a child. When a child is little its senses are tender. It sees, touches, feels, hears, smells keenly — when truth and love intermingle with this fine sensitivity a miracle occurs.

Ganesh through his drawings really lives with the animate and the inanimate beings in the story. This reveals not just inquisitiveness but an invisible order of a human being: it has in it the root of simplicity, the roots of distilled human nature, the roots of austerity, and the roots of aesthetics. This shows a humane approach to life.

The Holistic Approach

When I see a child playing, she plays with the material around. She makes a house, a mountain, a temple, and so on. And along with it, she sings, enacts it.

A girl does *Gauri-vrata* at the age of four or six or eight years. What she actually does is become the goddess herself. That is her first act of dressing well, fasting, eating a feast, wearing the best jewellery that she can; she sings, she dances merrily and all that. For her it is the first important celebration in her life. She knows the story of Gauri; her mother has told her about it. She begins from here on her preparation for her marriage. She grows sprouts, symbolising fertility, and she and her friends go to the temple. They derive great pleasure in doing this ritual act — lessons for life, the *samskaras*.

These and many others are lessons. They build the character of the person. Every such event has great significance, provides for the knowledge that we have, everything coincides somewhere and becomes part of the entire event of life. Like in tribal or folk people, one sees visuals like music, dance, drama, feasts — everything coincides.

In schools there is compartmentalisation of subjects into what is important and what is not. Art, play, library, excursions, all are considered subsidiary. This only isolates experiences further. It reduces the ability to extrapolate or draw from various experiences. Once I did a workshop with National Institute of Design (NID) students. We decided that we would pick on one theme and understand it in its holistic perspective. We thought we would work on the topic of hands. This was so because we felt that oftentimes we do not understand the significance of our hands in its fullness.

- First we did various exercises with the hand; this was like a warming up session.
- We would have everyday a 'resource person' coming into tell us the significance of hands in their lives. We had an architect, a dancer, a gardener, a tabla player so on. Everyday we would spend the first ten minutes with one such person.

done through play. It gives stimulus to the body, mind and heart. But in our system of education we consider play something negative. Actually play not only adds by itself it can also become a great source of knowledge in any subject. Here the teacher has to be a great master — only then will it be possible. Generally one feels that knowledge can come only through books. Normally people say, 'this child is playful and not serious'.

Once I was asked to do a programme on creativity for television. I had about twelve students between six and twelve years of age to join me. Some were school-going, some were not. I had asked them to get me any child they came across. We had a driver's son, a farmer's daughter, and a producer's daughter in the group.

A team of television personnel came to fix the backdrop, the platform, etc. I told them that we needed very little and that we would call them later. I also told them that we did not need ten to twelve people to do it and that just two or three would be enough. They were a little angry, but they had to accept. About the materials to be used, I told them that we would make use of whatever was available and that they should not make any purchases. So we worked with some pastel colours and water colours, ink, and some very ordinary paper. First I talked to the children briefly about what we were going to do. I had total faith in the children. We selected five elements: an old mother; a fairy; an animal; a tree; and water.

I then asked them to make a story out of these elements. All of them made their own stories. Then they had to select the story they liked the most. We decided that with couplets the stories would be even more interesting. They agreed and the next day their story had couplets as well. In the meantime I had asked the carpenters to fix cardboard boards and brown paper on the backdrop. On the backdrop each one of the children was given space and they were asked to paint the whole story which we had selected. Lo and behold, there was a big and colourful mural created in the whole process.

Now I asked the children to make a play of their story. They were told that they should stand exactly where each element was painted. So with the painting behind them, the children enacted the story. The whole exercise took five days. I did not add a word of mine. We got stories, paintings, sculptures, couplets and a play with a background mural, everything done by them.

The Environment

The environment or *prakriti* has many lessons to offer. One cannot ignore one's roots. They do have a definite role to play.

Once in Kerala we were conducting a workshop with children from the *panchayat* school. During lunch hour the children were eating their food on plantain leaves and when throwing them away, they made a mess in and around the waste-paper basket. I told the children to create something out of the leaves on which they were eating to express their skill and creativity. Then on another day I saw the children did not do it. I asked the children why this was so. One of the boys said, 'the teacher said not to'. And then I realised what the problem was. It was considered dirty to work with the plantain leaves on which they had eaten. The next day I asked them if they could just make something and throw it into the basket and decided that I would sit near the basket where they threw the leaves. They agreed. I sat near the basket and picked out about ten of these works and took them with me to the session. The other ones were thrown away. I carried these to the session. On my way to the class the children requested me to let them carry what they had made out of the plantain leaves. The session then began. I told these ten children each to pick up their own pieces and give little talks with them. They did it and they were truly marvellous. The next day I asked them to pick leaves fallen on the ground. I forgot to tell them to make something out of it. I did not understand their language and they did not understand mine. They assumed I must have asked them to make something and ended up creating beautiful things like birds, animals, flowers and bowls with these leaves.

It is easy to learn and communicate through one's own environment. While at Rishi Valley I asked the children to name the trees which they come across on their way to school. And up came as many as forty names on the board.

Learning by Doing

Let us talk of skill. Of specific skills, say weaving. When we talk of weaving, Gandhi comes to mind. In his programme of basic education Gandhi had introduced three hours of skill training such as agriculture, carpentry, weaving, and three hours of learning from theory. But even now the weaver in our country is treated as an untouchable. So we have not been able to carry forward this basic education.

Gandhi introduced weaving in the education system, not so that everyone should be called an untouchable. He in fact wanted us to understand the great skill of weavers like Kabir to know their minds. But disrespect for labour in our country has made our education lame. When there is so much work of skill all around us, why is it that our children don't get to see it or learn it? How did the potter learn his skill? We cannot take him to a class to impart his skill if he does not have a 4-year degree. What about the 400-year certificate that he carries with him?

Millions of such skilled people lie in a sorry state today. They are not given due respect

A Kutchi woman embroidering a textile for her child

A child doing tie-dye work

Drawing a lullby song. Artist: Ganesh

A painting. Artist: Ganesh's wife

Painting the mother goddess.
Artist: Jagdish

The mother goddess

Mural painting. Artist: Ganesh

Playing a stringed instrument

and that is how our educational system becomes disabled.

Sound

'Sound' and 'silence' are two very important elements in life, but that does not get noted in any lessons or courses. We have a word for children, '*Chup*!', meaning 'Be quiet!'. Here the teacher becomes a military person. Another thing that one gets to hear is '*Adab Palathi*' — meaning 'fold your hands together and sit properly, cross-legged'. You ask the child to close his mind, in effect, by asking him to keep quiet and stand to attention. Silence is a beautiful thing. In my experience children understand it very well. Prakash, Ganesh's son, can be silent for hours together just drawing mountains, rivers, snakes, birds and trees. Every child has the ability to make a song in the language he or she is most fluent in.

Real Learners or Mere Robots

All things that children acquire from books or good teachers — what we call knowledge — are only tools to become a learned person. If we assume these tools to be our ultimate goal in teaching or imparting knowledge, then it is but a dry exercise and we are reduced to being robots. The world of character-building, to be with oneself, to be true to oneself, all goes untouched.

It is only when you are constantly inquiring, constantly observing, constantly learning that you find truth, God or love; and you cannot inquire, you cannot observe, learn, you cannot be deeply aware if you are afraid. So the function of education surely is to eradicate, inwardly and outwardly, this fear that destroys human relationships and love.[6]

Notes

1. *The Collected works of Mahatma Gandhi*, Vols. XLVII, L and XXXVII, Government of India.
2. *Ibid.*
3. *Ibid.*
4. *Ibid.*
5. *Think of these things*, J. Krishnamurti, edited by D. Rajagopalan, Harper Perennial, 1964.
6. *Ibid.*

6

My Experiments with Education

D. Patnaik

As a village without food is of no use, a well without water is useless, offering of *ghee* to ash without fire is futile, similarly offering a gift to an uneducated brahmana is unproductive.

Experiments with Paintings and Plays

After my graduation in painting from the Indian Art School, Calcutta, I joined as an art teacher in a high school in Keonjhar Garh, the capital of Keonjhar, a feudal state in Orissa, in 1946. My becoming an artist was natural, but my becoming an art teacher in a high school was accidental. What has the art teacher to teach? A few drawings of fruits and foliage on the blackboard, which the students copy in their drawing books. The teacher guides the students to draw and colour some flowers and foliage from nature, no correction work, no preparation of lessons; enough time, sufficient leisure.

Teaching of painting in school has limited scope. In a week each class is allotted only two periods, each period of 45 minutes' duration. Besides copying they get the chance to draw from memory. By copying they learn to handle pencil, develop the sense of measurement through the eyes, not by using instruments, and develop a sense of proportion. Memory drawings help their memories to develop. When they are asked to draw from memory their brains start working. Their favourite and familiar objects appear before their mind's eye. A hut, a tree, a road leading to the hut. The sun rising behind a mountain, etc., are familiar subjects. As they grow in age, subjects of action, like working people, people carrying wood, ladies carrying water jars, grazing animals, occupy their minds. When they grow still more they draw multiple figures like markets, dance, snake-charmers, street singers and beggars. For this they get no encouragement from their parents. A good essay from memory fetches more appreciation than a painting from memory. One fetches marks, the other does not, though both need exercise of mind.

I encouraged the students who took an interest in painting by collecting their work and selecting and exhibiting it in the reading room. I prepared albums with their best paintings and named the collection *Rupa-Ranga*. These albums got a place in the reading room along with newspapers and magazines. Thus they were encouraged to paint more.

I took charge of the school garden. I wanted that there should be flowers. In the drawing period students used to go to the garden to draw and paint flowers in their drawing books. Thereby they came to love flowers and the garden too.

Forty years ago there was no hue and cry about ecology. In Orissa students observe two festivals, *Ganesh Chaturthi* and *Vasant Panchami*. They decorate the entrance gate and the *puja* platform with flowers and foliage. Generally mango leaves are suitable for the decoration. *Vasant Panchami* comes in early spring, and that is the time when mango trees bear flowers. The students out of ignorance collect budding twigs in plenty. I forbade them to do that sort of foolish act. I told them how they destroyed the mangoes by spoiling the mango flowers. I taught them to decorate using paper and cloth, not leaves. At that time there were a few schools; now their number has increased manifold. If the students are not checked from destructive practices, how can we save the fruit-producing trees from devastation? Our teachers should educate our students in this light.

Along with drawing and painting there was provision for teaching crafts such as carpentry, sewing and book-binding. I added clay modelling. Students selected subjects according to their taste. The handicrafts period gave pleasure to the students after boring bookish study. They enjoyed it when they created something out of wood, clay, cloth or paper. They understood the dignity of labour. If they take any one of the crafts as a hobby in their life, they may relax after the fatigue of mental work.

Many students took pleasure in clay modelling. I taught them to make human heads and figures. I started making images of Lord Ganesh and Mother Sarasvati. They joined me, and some of them learnt to prepare idols independently. They could get both money and appreciation. A student in those days, now a doctor of medicine, told me how the craft training helps him even today to relax. Another student, now a horticulture officer, said how his taste for gardening developed from his schooldays. Another student, now a handicraft superviser, remembered his schooldays when he learnt image-making, and how that training helped him in his career.

I found no teacher discussing subjects pertaining to students' development and character building. There was no discussion even of our great men, who had certain values and ideals. After school hours teachers exhaust their remaining energy in privately teaching those students who can afford to pay. They do not get time either to prepare lessons for the next day or to collect any new things for the students. I have never heard a teacher suggesting a good book to students from which they would learn things to build their character or enhance their knowledge.

A school makes a name by producing students who secure high marks in a Board Examination. Its fame depends on the number of students who pass the examinations. In this result-oriented formal teaching, where is the place for moral teaching? A student is recognised as good or bad from the marks he or she obtains, not from his or her character, which is not so transparent.

I do not know why the idea of building character in children keeps bothering me. I want students to be obedient to their parents and teachers. They should be kind to the poor, sympathetic to the needy; they should not adopt unfair means to pass examinations, should not cheat others, should not be selfish, should mind the development of their villagers, should love their motherland, and learn many other good things.

I started a weekly literary meet named Sahitya Sabha. Students who had an inclination for literature joined. I invited teachers to contribute articles, poems, stories, etc. That ran well. Some students got the encouragement to express their thoughts through literature. I was successful in publishing a biannual school magazine, *Vikash.* The selected articles, read in the Sahitya Sabha, got a place in it.

Then I started a dramatics club for students. There were certain occasions in school, like *Sarasvati Puja, Ganesh Puja,* prize distribution, where students gave entertainment programmes, like a few songs, recitals, one or two dances, some caricatures. I experimented with staging plays in place of a variety show.

In my one-act or two-act plays I wanted to demonstrate the teacher at his or her best. Stories of great men often inspire children's imagination. So I presented a play about Gopabandhu Das, the most beloved leader of Orissa. To me no such man of high character had ever been born in Orissa. His service to the poor and his sacrifice for the country remain matchless.

My second play was 'Master'. This was a teacher who wanted to do some reform work in a village, united like-minded young men of the village and advanced with his mission. People who were exploiting the villagers stood in his way. The teacher went on a hunger strike, and at last won the heart of the village people. I wanted to give the example of a sincere teacher who won not only the hearts of his pupils but of the people of the entire village.

Then came 'Ankura' (seedling), with much fun and frolic, to the stage. Children join the school with their own *samskaras*. Here they get the incentive from the teachers to grow their inner aptitudes. The teacher encourages the students to go on their own way and suggests that they be honest and sincere in their thinking and action. After passing from school and college, some become leaders, some artists, some scientists, some actors, some poets, some pleaders, some musclemen. In their own fields of work they do not forget their teachers' words or *guru-mantra* for the betterment of the people and the country. Even the muscleman saves weak people from the clutches of the wicked whenever he gets a chance.

In 'Achhuan' (untouchable), a veteran Gandhian tries to change the minds of the people of his village towards untouchability. He faces many obstacles; still he persists and fights alone till his death.

In this way I wrote ten plays to educate the school students. The child is always active. In education we are not dealing with passive material which we can mould to our purpose as a modeller moulds clay. We deal with active, volitional, purposive entities, each of which must develop along his own particular line. We can but guide the child.

The more friendly the relationship between teacher and pupil, the greater the potential for suggestion. If we wish our pupils to accept suggestions from us and to imitate us we have to love them and cause them to love us.

Let me narrate my experience. In Keonjhar Garh, in the month of *Kartik* (Oct.-Nov.) every year, a public function is arranged. An open-air theatre (*yatra*) is performed at night. The drama takes place on a raised square platform. The audience sit around it. One side of it is occupied by orchestra artists. Those who sit on the orchestra side do not get a clear view of the actors. People from the town and rural areas assemble there to watch the theatre for four to five hours. The students of the local high school generally sit in a group. The day when this incident occurred, students getting no other side vacant, occupied the seats on the orchestra side. They demanded the shifting of the orchestra group to the other side. The management refused their demand. As a result, when the music started the students shouted in protest and made the music stop. After some time again the music started and again the students started shouting. That nuisance continued for half an hour. The audience got bored. Neither party wanted to submit. I was watching the state of affairs silently with the audience. At last, hesitantly, I stood up and shouting at the top of my voice said, "Boys, just behave, you are school students, don't be selfish. If you want to sit in group, sit there. Alternatively you all can be accommodated if you sit singly or in small groups at convenient places." To my joy they chose the second, and many dignitaries in the crowd accommodated them without hesitation.

The child takes moral attitudes from its parents and teachers. In a public function I met the District Magistrate, who had come to preside. We were talking with each other when his child, a student of class X, came and occupied a seat next to him. There was still time for the commencement of the function. He called the child and instructed him in a very lovely manner. First he directed him to do *pranam* to me, as I happened to be his teacher; and secondly he told him to go and sit on the mat spread before the dais where other students were sitting. He told him that the seats arranged on the dais were for the teachers and other dignitaries. The child calmly left and joined the students on the mat. I appreciated this way of teaching proper behaviour to children.

Experiments with Children's Literature

After working in Orissa for eleven years, I left the school and joined the Faculty of Fine Arts in Banaras Hindu University, in 1957. It was altogether a different atmosphere. There were no children, only grown-ups.

Here my writing took a different turn. I switched over to children's literature. I am no educationist, I have little knowledge of child psychology, but I ventured to write literature for children. I had been successful in writing children's drama, and now I started writing stories.

Whenever we call a man or woman we address him or her with an prefix or suffix to his or her name, like Ramu Chacha, Sita Bhabi, Bhaiya Gopal, Beta Kanheya, and so on. We do not address a blind man as 'O Blind' or any man with a dark complexion, 'O Kale'. We teach children to call even a servant 'Bhaiya'.

In my story 'Their New House' I have tried to give this sense. A girl and a boy, Tuan and Tuin, invited a dog to live with them and guard them from wild animals. They did not know where the dog lived. They met a squirrel and Tuan called her to know about the dog. He said:

With a band of white
and a band of black
on your back
Where do you run
O ugly creature?
stop and hark.

The squirrel did not care to stop and went on his way. Tuin said, "No, that is not the way to call somebody. Let me call and see". She called out:

O lady of the wood
How sweet and bright
You look with the gown
Black and white.
Please stop and see.
We want thee.

The squirrel stopped to hear them. Another day they went to invite the peacock to live with them and protect them from snakes. Tuan addressed him:

O, Peacock you hear,
What a slender neck you bear.
Snakes are there
Where we live.
Your help we want,
Come and give.

The peacock did not care to respond. Tuin then called:

O king of dancers,
How well you dance.
Snakes will flee
If you come once.
Have pity and try
otherwise we die.

The peacock became very glad with Tuin's words and assured them that it would come.

Let us take the story *Gotia Thila Pila*, 'There Was a Boy'. There was a child who always wanted to play, neveratt ended to his lessons, and became naughty and obstinate. One day he got the chance to meet the young of some animals in the jungle. He called them to play. They refused to play, saying that that was not the time for play. They said they were not playing then. What the boy thought to be play was not play. They said, 'This is how we learn, learn to collect food, learn to defend ourselves from the attack of enemies, so we learn high jump, long jump, climbing, swimming, running. Don't you learn?' The boy came back disappointed. He felt hungry, found covered pots with marks over them: 'eat' written on one and 'no' on the other. He could not read, so did not touch them. One of the young animals asked him to read and to open the correct pot. The boy opened the wrong one and put himself to trouble. He came back, and from that day joined his friends and gave mind to lessons.

These are for children of the age group 5-8. For the age group 9-14 I tried to give some idea of vice and virtue, patriotism, self-sacrifice, etc.

In *Satya Bara Sata Papa*, 'The Banyan Tree of the Age of Honesty and Seven Sinners', there is a very big and old banyan tree in a village. Nobody knows its age. Some say it is of *Satya Yuga*, some say of *Treta*, some say of *Dvapara* and some say of *Kali*. Everybody tells a story to strengthen his hypothetical statement. When that tree is decaying and facing death, the villagers, being worried about the tree, start enquiring about the cause of the decay. They ask a philanthropist, a poet, a philosopher, a *fakir*, a *sadhu*, a wizard and an editor of a newspaper, and also approach the government. All say that a sinner has touched the tree and that that has made the tree die. Each one gives his own concept of sin and sinner. An ungrateful man, a greedy fellow, an exploiter of poor labourers, a corrupt public worker, a destroyer of communal harmony, a traitor to society and a man who extracts more milk from the cow, depriving the calf of its share are the sinners, they says. Each man named one sinner.

In my book *Phula Gachhati Mara Mara* 'The Dying Flower Plant' I want to show how corruption strides from top to the bottom. There was a king who ruled his state well. His subjects were happy and content. One day he felt something wrong with the people. They seemed unhappy and discontented. The king summoned his officers and ordered them to find out the cause of the change. Wise men, experts in social science, philosophers, thinkers, economists were invited to give their views. But nobody's verdict could satisfy the king. The courtiers, out of disgust, brought before the king a man, who by his outward appearance looked mad but whose eyes were glittering. Sometimes he smiled and sometimes looked grave. He was uttering some words which were barely audible and ambiguous. But his look created nervousness in others' minds. The king and his men could catch only a few words he uttered:

> For the sake of your child
> You left the cows wild
> And the flower plant died.

Neither the king nor his wise men could make anything of those words. Then the king went to an old lady who had once saved him by giving him shelter. She heard everything from the king and said, "O king, that man refers to you. You perhaps did something in favour of your son or near and dear ones, breaking the law of the land. Your officers followed in your footsteps, the common men then followed them. You lost the moral courage to check them and the result is before you". The king asked how could she know this from a few broken words of the mad man. She said, "We have one old story to make us conscious of our duty. There was a housewife whose responsibility it was to supply food to the cowherd in time so he would look after the cows. One day her baby cried, and forgetting her duty she went to it. As a result, the cowherd did not watch the cows and the cows destroyed the plant". The king then corrected himself and his officials.

I wrote poems with the hope that they would change the minds of children and they would become kind to animals. I do not know how far my voice will reach. If it reaches them, will the children act on them? If my stories move their hearts, when they will get the power of speech will they try to eradicate the cruel and inhuman activities now prevailing in society?

The bane of our educational system is that it is elitist and class-biased. At the same time it is unrelated to ground reality. As moral/ethical instruction and precept has no place in the educational curiculum, it develops a parasitic culture. Being divorced from ground reality, it produces swarms of ill-educated degree-holders who are mindless and directionless.

Who will give direction to these directionless students? A good teacher is always ready to take the risk of guiding such students.

7

Art as a Tool for Cultural Rejuvenation

Dinanath Pathy

CULTURE and education are complementary, inclusive of each other's essential ingredients. Culture paves the way for education and education is responsible for the flowering of cultural values in life. Life only survives in a balanced ecological condition. This interlinkage between culture, education and ecology is the very essence of life, its existence and continuity.

The interdependence which binds education, culture and ecology in an unending invisible thread is seen and experienced in human endeavour. Beneath this visible world is the inner perception of the universal life system, which is fundamental to all cultures across the globe.

The physical beauty, material culture, the abundance of variegated life manifestations, the visible cosmic order and chaos, all stem from the inner force which holds and sustains. Efforts to perceive are not instantaneous but a continuous process of unfolding and arriving at fundamental principles.

This journey could be termed the process of education, and the realisations gained from it are the spectrum of culture. The journey is performed in a visual world of sensory experiences. A successful journey endows a person with refined sensibility and enhances quality of life.

The journey is of an exploratory nature, making one understand forms, shapes, colours, musical sounds, rhythms and the inner harmony — not only of outer nature but of one's own physical and mental bodies. The journey begins in the womb and ends with death. It links a person with family, society, country, and the world at large, in an established cultural context.

I like to bring in 'art' as a tool to experience the aesthetics of this long journey. 'Art' not only as a skill as the 'art of living' in the present-day context, but as an 'act of transforming', where culture and ecology are relevant. I introduce 'art' not merely as a tool giving rise to consumer products but one which opens up the gates of a wider vision, a supramental consciousness of beauty and inner perception of a world order.

Art which breeds in creative and contemplative vision is a reality when translated into properties of culture and education. The cultural translatability of 'art' should form a component of our educational system and this should have meaning in the context of education, culture and ecology. The cultural translatability needs a language to transform a multilingual and multi-peopled phenomenon into a global cultural ethic. The present educational system should be able to provide this 'tool', the language of 'art'. This may be experimented with using a scientific temper as an alternative mode.

I illustrate below three experiments I had the occasion to carry out here and in Switzerland.

First Experiment

I was teaching drawing and painting to children from K.G. to Class X in the Kendriya Vidyalaya, Bhubaneswar. The school had no fixed syllabus to teach art. I did not want to provide model drawings on the blackboard to students for skill-oriented exercises. This I considered quite detrimental to the growth of creativity in children. On my initiative, the school provided sketch books to children and I inspired them to draw whenever and whatever they felt like recording from life experiences. At regular intervals I glanced through their sketch books and picked up sketches which attracted me from the point of view of innovative approach, creative excellence and pedagogic linkages. The subject-matter children drew in their sketch books was quite varied, with motifs from daily life and school books. They were attracted equally by a bicycle rider and the Prime Minister flying in a helicopter, the village goddess with protruding tongue, as well as their favourite film stars. The renderings of children varied a great deal depending on their faculties. These sketches also reflected their social consciousness and their interaction with their environment.

My intention in teaching visual art in school was to integrate it with the other subject areas a child is expected to learn, and not as an independent compartmentalised subject. This method yielded a lot of benefits. While making a picture a student used to learn not only about the picture he was drawing but several other facts and incidents connected with that picture and the entire cultural context. 'Art' in school therefore was a part of the total learning system meant to provide an aesthetic orientation to the child, whether it was in mathematics or in science, geography or literature. The Kendriya Vidyalayas project multilingual and multicultural content, since their students are drawn from all over the country.

Once while discussing with students the composition of a winter night, a number of possibilities came up. Since the students had come from various socio-economic backgrounds, they had different notions of a winter night. Some suggested a winter night in a sleepy tiny village around an open fire. Others imagined the winter night inside a house near the fireplace in the company of family members. A group of other students went for a more sophisticated environment and visualised the winter night warmed by an electric heater. When the pictures were drawn there were a number of innovative depictions drawn from different socio-cultural settings.

Second Experiment

In the early 1970s I had the occasion to teach visual art in a Swiss school. The idea of teaching no doubt was exciting, but I was not conversant with the language.

I thought about the problem of communication and rediscovered that the visual language needed to teach art is universal and can overcome barriers of language. At the end of the day I had a sense of achievement. I could make the class lively virtually without uttering a word. The visual symbols were enough to transmit the ideas of a multicultural set-up.

Let me elaborate on the symbols I used. At the outset I drew a conceptual world map to locate India and Switzerland and gave the children an idea of distance and direction. Within India, I focused on Orissa. The Indian and Swiss national flags gave the required identities to the geographical locations. I then drew a schematic map of an Orissan village, with the main street running east to west and other streets branching off like veins and veinlets in a human body. The temple, pond, well, school, the river, the distant hills, the mango grove, the coconut trees, the cows, goats, and chickens added to the beauty of the village.

Pointing to the typical house plan, with the cowshed at the front followed by the sitting room, verandah, open yard, sleeping room, store and kitchen, I explained the concept of the house and the joint family. The entire family sleeps in one room — something of a dream for Swiss children. Water is drawn from a well, filled in brass pots and carried home balanced on head or hips. Their eyes glowed with amazement. They rushed to me with their sketch books for a 'Frau'. The cultural symbols which I could construct helped in communication. The idea of a 'Frau' balancing a pitcher filled with water on her head, the other one on the hip supported by the right hand, and in the left hand a bucket, was most striking.

The other symbol was of a family with portraits, of father, mother, brothers, sisters, in their typical dress, ornaments and hairstyles. The Indian features came out sharply. I had a fruitful day in the school, visiting classes to give them the idea of an Indian village, family, specially the 'Frau'. The teaching was made exhilarating with singing, dancing and sharing one another's jokes and experiences.

During lunch break the students invited me to share their food. Some of them went home and brought for me a large cake with tiny Swiss and Indian national flags. This was a moment of great pride and excitement for all of us.

After this successful experiment I felt quite confident to provide the Swiss children an alternative to make them not only aware of but interested in India's socio-cultural traditions through visual symbols. Later, I illustrated a children's book, *Gita and Her Village in India.* The story of this book was provided by Eberhard Fischer and his wife Barbara. This book was basically meant for Swiss and European children to understand Indian village life through visual symbols. I illustrated how a small girl, Gita, spends a day in her village. The visuals spanned a wide range of incidents and situations from house interiors to fields, river fronts, the well, school, market and temple complex. It tried to provide a visual journey through an Indian village.

Third Experiment

In collaboration with my colleagues Eberhard and Barbara Fischer, I was associated with another interesting educational programme — popularising Odissi dance through visual symbols. This is yet another experiment with far-reaching significance. This project was sponsored by Unicef, Switzerland. The project consisted of an illustrated book entitled *Gita will become a dancer* and a kit with *ghoongur* bells and a few ornaments to put on while dancing. The background story on which the book was based was the life of a small girl who was inspired to become a dancer after watching the eminent Odissi dancer, Sanjukta Panigrahy, performing in her village. The story projected the determination of the girl and how she achieved her objective through sheer perseverance. The visuals of the book as well as its story content are interesting enough and informative enough to teach a child how to dance at least for five minutes.

These experiments had wonderful results. The challenges that face our educational system are stupendous. Ways and means must be devised to tackle them. I have designed two modules which could be tried out in Indian schools. The implementation does not call for extra cost. It only needs a reoriented approach for re-structuring our conceptual framework.

Module 1: Art at the centre . . . . contextual linkages in which art/art educator plays a vital role.

Module 2: Child at the centre — cultural dimensions of learning, challenges of explorations, changes and socio-cultural identity.

A separate syllabus is not required to teach art. Art cannot and should not be taught in classroom situations. Art should be a binding medium holding together the total teaching curriculum, reinforcing, permeating and enriching the educational structure both at home and in schools. Art has refreshing and innovation-inducing qualities and it can bring about a total change at the perceptual and working levels.

To conclude, I may again emphasise that art is the reflection of the universal order. It is the visual manifestation of the invisible spirit. Art is not the negation of science, technology, and modern living. It is a rejuvenating tool.

8

Photography in Education

Ravi Chopra

EDUCATION has continued to evolve, diversify and extend its reach and coverage since the dawn of human history. Every country develops its system of education to express and promote its unique socio-cultural identity and also to meet the challenges of the times.

The National Educational Policy marked a significant step in the history of education in independent India. It aimed to promote national progress, a sense of common citizenship and culture and to strengthen national integration. It laid stress on the need for a radical reconstruction of the educational system to improve its quality at all stages, and gave much greater attention to science and technology, the cultivation of moral values and a closer relation between education and the lives of people.

The introduction of systematic, well-planned and rigorously implemented programmes of vocational education is crucial in the proposed education reorganisation.

There are various methods for making the curriculum more interesting, enjoyable and child-centred. Attempts have been made using media like puppetry, theatre, songs, games, etc. In this direction I have used photography to make subjects like the environment, heritage, nature, people, festivals, places, etc., more interesting and meaningful for children belonging to rural areas, slums, streets, schools and tribal areas.

Photography mirrors reality, an art as well as science; it captures and reproduces reality with accuracy. Its services span the range of human needs, both practical and emotional. It plays a vital role in helping to preserve records of different fields of study, in preserving the cultural heritage, in business, industry, education, medicine, criminology and defence services, coupled with keeping a record of special occasions in day-to-day life.

In children, much like in adults, photography provides the thrill of catching and freezing a moment, a scene, a smile, a movement and looking back at it. It is a creative art involving the child physically and psychologically. The pride of reproducing a moment encourages the child to look at the world through his own eyes.

Communicating information without distortion is always difficult. Everyone approaches a new subject with a different and personal frame of reference. The special virtue of photography is that it eliminates ambiguity, allowing your message to be received with maximum impact. Add photography to words and fewer words can be used to convey more and better information.

Photography helps make education more interesting for children. It does so directly through slides, movies and other visual aids.

Experiment in Photography

My involvement with children began four years ago when a little girl of seven, living in a slum area behind my house, approached me to have a peep through the viewfinder of my camera. The sheer ecstasy of the little girl gave a new dimension to me. I decided to work with underprivileged children from the streets, slums, villages, remote areas and tribal communities. Over the years my efforts began giving the handicapped and runaway children, living with very little resources, a new sense of confidence and a job-oriented skill which would improve the quality of their life, their personality and life-style. It has aroused the creative instinct in them, providing a chance to contribute positively to society.

For four years I conducted a series of workshops in and around Delhi. My aim was simple: to teach children the practical aspects of photography and its potential in tapping their creativity. The idea of working with underprivileged children appealed to me immensely as I firmly believed that with this training programme, at least some of the children might turn this knowledge into small commercial ventures, creating job and business opportunities. The interest thus aroused might perhaps lead to an increase in social awareness and raise the standard of photography in India. This project would also give young minds greater confidence in their work and lead to self-reliance.

The workshop gave me an opportunity to expose the children to diverse topics such as air and noise pollution, water pollution, poor sanitation, common diseases, malnutrition, land and soil, wildlife conservation, deforestation, environmental awareness and pride in national heritage and culture.

The spontaneous response of the children and the photographs taken by them, encouraged me to seriously develop this project with like-minded organisations and people wanting to work with the underprivileged. My attention was drawn towards working in remote areas like the villages of Himachal Pradesh where the children are innocent, the scenic beauty intact and untouched by the influence of modernity. Armed with Snapper pocket cameras, donated by Agfa-Gevaert, and with funds from UNICEF, I conducted the first photographic workshop in the village of Shalana.

Returning to Delhi with a new-found passion, I embarked on conducting further workshops on the New Delhi railway station platform, the slums of Govindpuri, Sangam Vihar, Raghubir Nagar, R.K. Puram and in schools run by the Delhi Administration. Through the World Wide Fund for Nature I also held workshops for the children of Sardar Patel Vidyalaya, Kendriya Vidyalaya and the Army Public School. I conducted workshops for tribal children and handicapped children of Maharogi Sewa Samiti (Anandwan, an *ashram* run by Baba Amte) in Maharashtra.

In the workshop for deaf and mute children, sign language was used for teaching. Baba Amte observed that since those children did not have much opportunity to express themselves, photography was the right medium for bringing out their creativity.

Methodology

Throughout my experience, it was the problem of survival of the slum children that preoccupied my mind. It was evident that they lacked the necessary inputs and the socio-economic environment to cope with the demands of the educational system. Their living conditions made them disinterested in studies, constantly driving them towards despair and frustration. To give direction to the life of these children, I established a society called Disha, which started a photographic training institute to train street and slum children.

The project also aimed to enhance the artistic bent of the children's minds and help in increasing their technical abilities. The use of the camera was encouraged not only as an instrument but as a means of acquiring a fresh perspective on the world around them. A child interested in making photography a career can venture into any branch of the art like commercial photography, institutional and specialised photography, photojournalism and press photography at a later stage. The main themes chosen for the workshop were:

1. Documenting urban life, monuments, nature, etc.
2. Environmental awareness
3. Imparting an artistic and aesthetic outlook
4. Vocational training

The project was initially discussed with voluntary organisations and professional photographers. Sponsors were located and target areas identified. The children who were to be involved in the workshops were chosen after judging their IQs, educational standards, etc. The work schedule was planned according to the availability of their time. The workshop dealt with the following topics:

1. Basic theoretical knowledge and practical training.
2. Darkroom work.
3. Picture composition.
4. Handling of different subjects.
5. Commercial application and picture presentation along with total personality development.

Unexpected results were noticed. The children took the project very seriously and showed keen interest in photography. They started seeing their surroundings with a new outlook, which generated confidence among them, created a feeling of creativity and strengthened their desire for self-expression. They went around with cameras and photographed whatever caught their fancy. This inculcated a relationship between them and their environment. The processes and the chemistry involved in developing, printing, enlarging and composition of photographs intrigued them. Children from different social backgrounds and economic status worked together in groups, thus enhancing the feeling of individual and community development. While photographing historic monuments they were linked to their past and developed a desire to conserve these for future generations.

The slide shows and documentaries shown to them helped them to understand their rights as human beings and their responsibility towards their environment. Visits to newspaper establishments gave practical demonstrations of the actual use of the visual medium.

There were difficulties encountered during the project as well. I had to look for cameras, which were ultimately provided by a reputed photo goods manufacturing company on a returnable basis. I had to arrange for the equipment and chemicals to set up a workable darkroom.

Case Studies

WITH CHILDREN FROM SLUMS

Six workshops were organised in the Govindpuri slum of South Delhi, in which I trained 96 children. Govindpuri is situated about 20 km from downtown Delhi. It is the second largest slum area in India after Dharavi in Bombay. In its vast number of *jhuggis*, some 35,000 people live. Children in the age group of 12-18 years were given basic knowledge of photography and its practical aspects.

The initial week was spent in teaching the basics and theory. The next week was devoted entirely to the working of the camera and its functions. This was followed by picture presentation and the commercial aspects of photography. Then came the practical sessions, where children started taking photographs and were taken on photo excursions to various sites.

An exhibition of photographs taken by these children during training was held in Govindpuri itself for the benefit of the residents and the guardians of the children. The exhibition drew a large number of visitors. Many parents from the adjoining *jhuggis* approached me to conduct similar workshops for their children. It had obviously aroused a lot of interest in the media, which was amply proved by the coverage in almost all the national dailies.

This was followed by a two-week refresher course for trainees on the basis of their interest. On seeing the response of the children the local NGO also funded a darkroom in Govindpuri for the benefit and use of these trainees under the guidance of Disha. It is now fully operational and is being used by the young shutterbugs. A complex was provided to Disha by the local government in the Ekta Vihar slum cluster in South Delhi to conduct vocational training courses for youngsters living in the nearby slums. This programme was started in January 1994 and was funded by UNICEF.

WITH SCHOOL CHILDREN

Children from the Army Public School, Sardar Patel Vidyalaya and Kendriya Vidyalaya in Delhi were involved in workshops of two weeks' duration each in which they were given environmental awareness through photography. This was sponsored by the World Wide Fund for Nature, India.

The children took this project very seriously and showed keen interest in learning and understanding all the aspects of photography. During this training they started seeing their surroundings with a new, involved outlook which generated confidence in them and resulted in strengthening their desire for self-expression. The relationship between them and the environment manifested itself in various ways.

The experiment amply proved that children possess a unique faculty of expression and creativity and only need the right guidance and encouragement.

Success Stories

1. Virender and Ramesh, who had run away from home in U.P. two years earlier and

lived on the New Delhi railway platform working as coolies, are now working as darkroom assistants in a studio in Paharganj and also freelancing in photography. Their lives have become meaningful and happy. Ramesh says, "If I had not met 'Sir' at the railway platform, I could be back at a corner somewhere selling drugs or killing somebody for money".

2. The photographs taken during workshops were displayed in an exhibition by schoolchildren in their schools. Photographs were used in school magazines. Twelve selected photographs were sent by the United National Information Centre, Delhi, to be displayed as resource material in a conference at Bogota, Colombia, by the United Nations Environment Programme.

3. Krishna, who lives in the Shastri Market slum in R.K. Puram, was employed as a photographic instructor in Hansraj Model School in Delhi after the completion of the workshop. In the evenings she sells vegetables in the main market.

4. Horilal, from the same *basti*, has opened his own studio in the Moti Bagh area. He was earlier working as a *kabadiwala*.

5. Tulsi Ram, Tofeeq Ahmed and Rajpal, belonging to the Dr Ambedkar slum *basti*, have been assigned to make photographic case studies of different types of low-income urban settlements by USAID and UNICEF. Various agencies like the Aga Khan Foundation and Navjyoti have engaged their services from time to time.

6. The trainees from the Sangam Vihar resettlement colony and the Govindpuri slum area have formed three cooperative groups to cover the various events in their localities.

7. Exhibitions of photographs taken by children have been sponsored by the Delhi Administration, UNICEF and IGNCA from time to time in Delhi.

8. An exhibition on the socio-economic condition of the Aravalli hills was developed and designed by the trainees of the R.K. Puram slum *basti* with Disha and rotated among schools and colleges in Delhi. This was widely acclaimed by the press and the public.

Conclusion

My experiments with children from different backgrounds proves that photography can become a rewarding and enriching method of learning. It also keeps children occupied with an a vocation while contributing to their creative expression, self-confidence and employment potential.

9

Education for Value-Creation and Leadership

A Case Study of the Rangaprabhat Centre

N. Radhakrishnan

EXPERIMENTS require courage, conviction and great application of mind on the part of those who undertake them, and if they are related to children's education, they should be characterised by a child's sensitivity and imagination. Adult society imposes its parameters on the child and thinks that in conformity with the fast-changing social milieu the child should be offered a variety of learning materials to help it reach the top. The tragedy in this is that we do not care for the **child** in the child. Wordsworth's assertion that the child is the father of the man seems to have been forgotten by educational planners and experts.

The first ten years of schooling should enable the child to extend his frontiers of creativity, imagination, self-confidence and character formation, thereby helping him to face the problems of growing up. The aim should be the enrichment of the consciousness of the child, in no way clouded by the obsession of offering knowledge and skills to earn a livelihood. Mahatma Gandhi offered one of the most sound frameworks in this direction when he insisted that the kind of education we might think of for the future should be one which would help the child to hear the music and harmony of nature and would draw out the best in him.

A highly disturbing aspect of modern education, in sharp contrast to the ancient system, is the manner in which schools in far-flung rural areas are being organised. It appears that most of the innovative methods or experiments or the thrusts and facilities do not percolate down to the rural schools which, through sheer neglect on the part of all concerned, have made been unattractive and uninspiring. There is a lot of unlearning taking place in these so-called centres of learning.

This brief discussion leads us to an experiment undertaken by a group of artists and educators in Kerala under the banner 'Rangaprabhat' to offer a support system of value creation in children and prevent school drop-outs through drama and a host of other activities. Though this case study will describe the salient features of the general experiment, the emphasis of this discussion will be the nature, scope and impact of three specific programmes currently undertaken by this group: (1) the efforts made to weave indigenous cultural traditions, arts, music and crafts into the learning process; (2) intensive and self-generative programmes in forty select schools in the rural areas of Trivandrum District to create in each an atmosphere where both teaching and learning become relaxed and enjoyable; (3) a regular Saturday carnival for children from schools in the neighbourhood at which they get a series of opportunities to involve themselves in craft activities, dance, music, painting, creative dramatics, improvisations, make-up and oratory under the general supervision of highly qualified teachers and artists.

A group of artists and educators led by Professor G. Sankara Pillai, a great artist and visionary, and his disciple K. Kochunarayana Pillai, initiated about 25 years ago at a place called Venjarmood, about 20 km north of Trivandrum, what was at that time called an experiment in primary education by the integration of non-formal and formal education through the liberal and creative use of the vast opportunities offered by children's drama and by offering opportunities to selected children to spend their leisure time in activities that would enhance in them a spirit of cooperative endeavour, involvement in craft activities, creative dramatics, story-telling, improvisations, creative dance and folk music. The aim was to offer facilities and opportunities to children to supplement and augment their classroom learning in an atmosphere of relaxed freedom which would promote a creative involvement in them. This led to the gradual development of an alternative 'campus' or 'school' centred around the house and the thatched shed that was put up near the house of Sri Kochunarayana Pillai, himself a distinguished teacher at a local school.

The experiment very soon turned out to be a quite useful one and went a few steps ahead of the popular Western concept and practice of theatre-in-education. The manner in which the programmes developed at this centre indicated that it succeeded in (1) creating community consciousness; (2) exploring the moral, spiritual and cultural dimensions of education; (3) creating awareness among the parents of the locality of the need for letting children get involved in what are euphemistically described as non-academic programmes; (4) developing leadership qualities in children; (5) confidence-building; (6) character formation; (7) value creation; (8) making learning more enjoyable; (9) creating awareness of indigenous cultural traditions; (10) involving children and teachers in developing love, respect and pride for the folk arts of the country and encouraging their preservation.

The experiment began with story sessions through which both Professor Sankara Pillai and Kochunarayana Pillai were able to attract the attention and sustain the interest of a group of over sixty children initially. Story-telling gradually led to creative dramatics, during which the children improvised situations and characters. An attractive feature of this experiment was the time the children spent in games and songs. They were encouraged to develop dramatic moments and situations from their games and songs. The atmosphere was participatory and never competitive, in sharp contrast to the situation that existed in schools. On holidays the small house and shed, which had by now assumed the name 'Rangaprabhat', would be swarming with children of the age group between five and sixteen, involving themselves in activities of their choice under the guidance of elders. The emphasis was to encourage relaxed and participatory learning and acquisition of skills and confidence.

What guided all those who involved themselves in the experiment was the emphasis which Professor G. Sankara Pillai placed on **understanding** children and on **catching them young**, as well as on leading them not under compulsion but to learning through a variety of activities which are not available in the schools in which they study. There were those who could offer help in regular academic learning also, so that the children did not lag behind. In short, very soon the house of Sri Kochunarayana Pillai became the centre of new and bold experiments in non-formal education essentially at primary school level.

A monitoring group attached to this work indicated in the first two years four important aspects related to learning, which encouraged the organisers to go ahead with confidence. They were: (1) there were no drop-outs from among those who were coming to Rangaprabhat; (2) the activities offered by Rangaprabhat helped the children sustain interest in their academic programme; (3) the children scored more marks in their annual examinations; and (4) they exhibited more confidence in their ability to express themselves and displayed better hygienic and personal habits.

Professor Sankara Pillai, the brain behind this experiment, was encouraged to streamline the activities of Rangaprabhat by introducing puppetry and children's drama. He launched the revolutionary concept of children's drama in Malayalam for education, for which in the next 22 years he wrote several plays — all for the children of Rangaprabhat to enact. His aim was not to create new actors or actresses but to help children involve themselves in activities that would enable them to imbibe values and attitudes which the formal education set-up does not offer them at the moment. The emphasis was on helping children learn through creativity. As Professor Pillai would always say, **the child likes to learn but hates to be taught**.

The third important stage in the development of Rangaprabhat came when efforts were made to involve the local population in its activities so that it would truly become a community centre of non-formal education for children and would gradually begin to influence parents in the rural area to understand their children.

In collaboration with the Indira Gandhi National Centre for the Arts, it launched a **summer mobile awareness creation campaign through children's plays** all over Kerala over a period of 45 days last summer. The programme, a significant step insofar as its educational content was concerned, turned out to be a highly satisfying experience to both the organisers and the student participants. The programme, which covered a distance of over 2,000 km in fourteen districts of Kerala, had many tiers: play presentation, exhibitions and interaction with artists, seminars, visits to historically important places, and exposure of rural children to an exciting phase of group living.

Though the primary purpose of the programme was to demonstrate to educational experts the possibility of using children's drama for the propagation of values and ideals and character formation, the exercise was to offer the children, numbering forty, an opportunity to spend their summer vacation profitably. The programme had an interesting schedule. In the morning the children would arrange the exhibition and in the afternoon there were seminar and discussion sessions at which two or three persons associated with education or children's theatre or any other innovative activities would share their experiences. This would be followed by the play presentation. After the plays there were informal sessions of discussion with some senior and knowledgeable persons of the locality on the importance of the place, its literary connections, historical significance, eminent persons, folk arts and festivals, folk stories, games, and so on. Perhaps these sessions were the most important ones, since they offered the children a rare opportunity not only to learn of the rich heritage of the different places in their state but also to see leading artists, writers, educators, social activists, theatre personalities.

Another attractive aspect of the programme was the unmistakable stamp of the Rangaprabhat experiment in developing self-sufficiency and leadership qualities in children. The group did not have anybody other than the children and two or three senior functionaries during its 45-day tour, and the children had to arrange everything themselves wherever they went. From arranging the exhibitions to packing and loading, they had to do everything. There was no technician of any sort. The little make-up they had to do was also managed by them. Sri Kochunarayana Pillai and his wife accompanied them and watched them with amused delight and with a sense of satisfaction as they completed their journey. Twenty-five years of hard work had not gone waste.

The summer mobile awareness creation campaign turned out to be almost a passing-out parade or a graduation ceremony, and those of us who helped the children in this programme were gladdened by the success of the programme. The tremendous confidence and organisational skills they displayed was an eye-opener to the elders at the time of assessment of the programme.

Rangaprabhat has now become a community centre of extremely important educational experiments involving not only students and teachers but also parents. It has earned the distinction of acting like a bridge between the school and the community. This does not mean that the entire surrounding village community always appreciates what is going on in this centre. It has to weather the strident and orchestrated criticism of a community which entertains visions of all their children becoming doctors or engineers or professors and do not want their children to waste their time in drama activities which, according to the conventional opinion, are no good. The apathy of ordinary citizens towards creative involvement of their children is very strong and they feel that these are diversions and certainly will affect their children's efforts to secure more marks. It is this which Rangaprabhat has been fighting. It should be mentioned here that Professor Sankara Pillai, who had a vision and firm conviction of the extremely important role this experiment could play, was not ready for any compromise, for he always reminded us that we had nothing to compromise about:

> We are offering something new — a healthy alternative, however small it may be. The experiment aims at convincing those who are concerned with education that what we want is the realisation of the simple truth that the child is to be respected and understood, and all our educational experiments should not forget this single basic aspect.

The Centre is in its silver jubilee year now and this year witnessed another landmark in its resolve to involve over forty schools in the rural areas in Trivandrum District in creating community consciousness and stressing the importance of moral, spiritual and cultural dimensions of education. This three-tier programme involves in its first stage identification of one teacher, two students and one representative of the local youth or art clubs near the school and getting these four from each school to the Rangaprabhat Centre for a ten-day orientation course. After the orientation programme when the teacher, the students and the youth club member are back in their schools, they try to create a nucleus for a variety of activities on the basis of the guidelines offered to them but also taking into account local response. One trained and qualified member of the Rangaprabhat Centre offers help in sharpening of the programmes undertaken at the school level. Once a month the child artists of Rangaprabhat visit each of the forty schools in groups and present one or more plays. Though this particular experiment began only four months ago, it looks promising and it is hoped that as the programme gathers momentum, more inputs will be added to this modest initiative.

The experiment is not stereotyped or static with set norms and limited vision. Contemporary issues and concerns which do not have political or religious overtones are woven into the activities of the group. Tolerance is an important virtue which the group tries to inculcate in the children, and this year being the Year of Tolerance, the group has been presenting a children's play *The Three Wise Birds*, written by the present author. The presentation of this play, besides helping both adults and children to cultivate this important virtue, has definitely added to the conviction of those children who have been part of this experiment.

10

Education Through Art

Nita Mathur

As empiricism, experimentation and demonstrability reign over the world order, there is a deepening silent crisis in education marked by eroding wisdom, depleting values and denuding self-knowledge. This crisis cannot be resolved by improving literacy rate figures, nor by making policy interventions, much less by creating data banks and building up information storehouses.

The word 'education' is often employed synonymously with literacy, particularly in bureaucratic and political parlance, as a kind of divine weapon — the *brahmastra* to combat all human problems. Material and human resources are hence directed towards positing and achieving formidable targets of total literacy. Education programmes lay emphasis on learning to read and write as also on specialising in various academic disciplines.

The processes of education place certainty and method before ingenuity and spontaneity; abstraction and categorisation before imagination and classification; information before knowledge; and intellect before wisdom. While this is conducive to the promotion of universalised information modules, it simultaneously relegates the cultural and local knowledge, skills and wisdom to the background. The 'educated' people learn to appreciate not so much the colours of the rainbow, the fragrance of a flower, nor the song of an illiterate woman, as the underlying scientific theories of reflection and refraction of light, liquidity, hydration and acoustics.

In the present social system marked by increasing fragmentation and divisiveness, the narrowed vision of education as literacy only reinforces centrifugal forces. Unlettered has come to be understood as being coterminous with uncivilised. A large section of people such as those engaged in weaving and sculpting, equipped with traditional knowledge and wisdom, are branded as 'uneducated', being given to oral rather than the textual tradition for transmission of skills and cultural heritage. The situation gets worsened when literacy is used as an indicator of development. Education in the limited sense of literacy *per se* is certainly not the means to realise cultural development which, in itself, according to Vatsyayan (1991), is the capacity of the individual to be in harmony within, and to create harmony and peace without. There is an urgent need to take cognizance of the situation and to evaluate and re-evaluate the pervading concept of education.

In its original sense, education is constituted of, among others, (i) *shiksha*, which means art; (ii) *adhyayana*, which means to turn the mind towards, to observe, to understand; and (iii) *vinaya*, which means to lead out in a particular way. The three components together make for acquiring knowledge, inculcating a sense of social responsibility and, quite importantly, strengthening character. Students in the traditional milieu led a life of discipline and austerity, upholding ideals and virtues. They derived inspiration from the life-style of their teachers, the processes of nature and implicit cosmological ordering.

The process of education began at the embryonic stage when hymns were recited, sacred verses were chanted and rituals were performed as part of pregnancy rites to instill morality, religiosity and spirituality. Das (1986) writes, the children were taken into the open to admire the gay flowers and green leaves. They breathed the pollen driven by wanton winds, witnessed the bright plumage of the dancing peacock and were treated to the music of the gurgling brook and the sweet songs of birds. Some of the centres of learning at later stages were the hermitages of *rishis* and other learned men. The children who lived with them shared the toils, studied under their tutelage and served them with much sincerity.

Knowledge was not a matter of mere memorisation and intellection. It was to be experienced, realised and lived. In the words of Annie Besant (Das, 1986:61-62):

> Not from them [the modern 'Civic Universities'] will come sublime philosophies or artistic masterpieces, but they will doubtless produce men of inventive genius, miracles of machinery, new ways of annihilating space. But in a country in which a man is valued for what he is, not for what he has, in which a man's life consisteth not in the abundance of the things he possesseth, the Indian ideal is the most suitable. The essence of that ideal is not the forest as such but the being in close touch with Nature; to let her harmonies permeate the consciousness and her calm soothe the restlessness of the mind. Hence it was the forest which best suited the type and the object of the instruction in the days which evolved *rsis*; instruction which aimed

> at profound rather than at swift and alert thought; which cared not for lucid exposition by the teacher, but presented to the pupil, a kernel of truth in a hard shell, which he must crack unassisted with his own strong teeth if he would enjoy the kernel . . . instruction which thought less of an accumulation of facts poured into the pupil's memory than of drawing out in him the faculty which could discover the truth, hidden beneath a mass of irrelevancies; of much fruitful study the Hindu Ashrama in the forest is the symbol.

This prepared them for earning a livelihood and largely for social and spiritual existence not only in the present but also in future births. Education thus has to be distinguished from literacy in being holistic in approach, enfolding the multiple dimensions of human living and their relations and interrelation with nature, environment and cosmology. Systems of traditional knowledge are broadly characterised by a distinct sense of oneness of human beings and the world around. The earth, water, fire, birds and animals are all suffused with consciousness which unites them with one another and with human beings.

Knowledge such as this is experiential. It develops from and survives on the cultural substratum of the people who create and use it. Thus education consists of beholding, receiving, embellishing and handing down traditional wisdom. A piece of Rabari embroidery, for instance, is also a kind of text, a visual text quite different from written text. The Madhubanis aspire to depict their legends, re-create their own lives and transcend worldly existence through their paintings. A painting is subtle and replete with meaning and message, carrying substance beyond the contours of figures and spread of colours. The patterns and motifs of trees, birds, fish and others, singular and in association, portray the gamut of beliefs, concepts and understandings, in essence, the life-style and world-view of the Madhubani people. Elders sit in the midst of children who assist them at various stages. The technique and the conceptual context gets passed on from one generation to the next through verbal articulations and depictions as part of collective social memory.

Conveyed by oratory, preserved by memory and transmitted by legacy, this system of education is related with the everyday life of a people, which makes it meaningful and fulfilling. It becomes a pathway to self-knowledge and truth. The many branches of knowledge and the arts share the cosmogonic base. As Saraswati (1994) has pointed out, the Rathvas and Saoras regard god as the first painter and, interestingly, the painting itself as god. The act of painting thus gets interpreted as 'reading' the god. The people receive inspiration for painting in dreams and in altered states of consciousness. While the content of Rathva paintings derives from the pictorial history of the universe — the work of the first painter — the subject-matter of Saora pictographs is life in the underworld as revealed to the priest in dreams and in a state of trance. The paintings, songs, dance, theatre — all employ the human body, sensibility and sensitivity to transcend its own limitations to achieve confluence with the Impersonal Principle. The artists, then, are *yogis* or *sadhakas* engaged in spiritual pursuits. The inner silence, contemplation and refinement transmute the outer chaos and noise into creative works. This makes for rhythm, harmony and orderliness in the world around.

The concept of art here is not confined to exotica. The artist is not just one of a select few constituting the elite. Each person, being equipped with faculties of expression and appreciation in one or other medium, is an artist. The free lines drawn in leisure by a child are as much a product of art as a portrait drawn by a serious painter; the *nautanki* of Uttar Pradesh and the rhythmic, co-ordinated patterned movements of village women are as much dance as a Bharatanatyam sequence performed by a celebrated dancer. The experience and participation herein are more important than the stylisation and perfection of the finished product. The issue is one of realising one's potential and developing it as an integrated aspect of growing up.

An essential requisite is the incorporation of the aesthetic dimension into education not as training in skills but as an agency for developing a synchronous, holistic life-style and perspective. The position of dance in this context is extremely important, both as a component of education and as a receptacle of the elements of education. The dance movements characterised by rhythm and symmetry are known to stimulate the various bodily systems, enhancing their power and efficiency. This is closely followed by the development of balance and proportion in the body. To children, dance unveils the many channels through which emotions and states of mind may be expressed. There are distinct gestures, postures and facial expressions that communicate the shades and intensities of *rasas* or inner states and aesthetic experience.

Dance performs the cathartic function of releasing pent-up emotions and drives. The control and discipline of the body so arduously acquired by children in dance is inseparable from that of the mind. In identifying with the enacted character and the situation of dance, children are lifted out of the disturbing unconscious realms of the mind. They employ various defence mechanisms — compensation, atonement, self-actualisation — in the dance situation as means to surmount worrying thoughts and muddled instincts. This prepares the mind for comprehending and retaining instruction in schools.

In addition to the impact on the bodies and minds of children, dance leads to the cultivation of perseverance, reverence and tolerance as cherished traits. Children dancing together are engaged in a group activity and begin to respect the diversity among them and distinctiveness of people. They get committed to social responsibility and obligation. They learn to organise the social space — merging with others in a uniform group while yet maintaining individual, personal niches. The habits, coping styles and behaviour developed in the early years largely last into adulthood.

More specifically, the earliest introduction of dance into the lives of Tamil brahmans is at the pre-natal stage, when pregnant women watch sequences of Bharatanatyam dance. Some of them told me that the song and rhythm in the dance delights not only them but also their forming babies. Indian mythology abounds in instances of learning in the womb. Some women carry children only few months old to dance performances. One of the chief reasons for exposing children to dance from the very beginning is to ensure in them the inculcation of interest in and appreciation of the art.

Children seek admission to dance schools on the festival of *Vijayadashami* — the fruition of the preceding nine-day rigorous worship of *devis* or feminine deities and the day commemorating the victory of Lord Rama (righteousness) over the demon Ravana (unrighteousness). They make offerings of coconut, betel leaves and nuts, bananas, lime and sweets to the teacher. After worshipping the teacher and the gods, they are initiated into the tradition of dance. Girls always pay homage to the earth before beginning to beat their feet on it and on withdrawing from it. The song and thematic content of the dance centre around episodes from mythology that establish the universality of divine consciousness in all objects and beings. Consequently, the earth, sky, water, stars, trees — everything that exists — throbs with sacred life. The entire cosmos is venerated.

The learning environment is one which is familiar to the children, one in which they feel rooted and responsive. Little dancers often address the male teachers as *tattha* and women teachers as *akkal*, which are also terms for father's father and father's sister respectively. They weave ties of kinship with other students and with their parents and relatives. The entire group, with the student and teacher at the core and others at the periphery, is knit together as a family in this way. Apart from throwing their bodies to rhythm and moving together, the children share the anxieties that fill their days and celebrate occasions of happiness. Inevitably, there are instances of rivalry and jealously. One of them may intentionally strike a hand against the arm of her neighbour, scorn a slow learner or hasten to point out the misstep of a partner. A teacher aspiring to build the character of children efficiently and effectively manages such situations, converting hostility into affection and competition into cooperation.

The teacher is not a distant figure. She disciplines the children, narrates stories to them and exhibits interest in all their activities including those in academic and familial dimensions. The children in turn confide in the teachers. A large number of them regard the teachers as role models. They observe the teachers closely and emulate their behaviour. Making the choice of a teacher — proficient and virtuous — is a difficult task for the parents, particularly because the dance curriculum consists of learning the repertoire of body movements, gestures, expressions and the assimilation of the underlying life philosophy in its entirety.

Children form ideas about the seen (*pratyaksha*) and the revealed (*paroksha*), about form (*rupa*) and the formless (*arupa*). This input makes for the percolation of spirituality, devotion and compassion. The reverberations of songs, music and rhythm of co-ordinated movements allay fears, insecurity and tensions. The emotional turmoil and conflict — the natural fall-out of growing up in an age of competition, strife and fragmentation — get aligned and appropriately channellised. The dispersion of mental preoccupations facilitates psychic integration, harmony and synchronisation.

Besides other things, song and dance lay out the course of 'right' behaviour. One of the songs that most children learn suggests that activities of art should be undertaken during morning and evening hours; blessings from gods and elders should be sought to achieve success in endeavours; honesty, compassion and courage should be foreseen and prepared for; and women should be simple, humble and modest. Another theme promoting higher ideals and values is one of Krishna and his poor friend and devotee, Sudama. The ragged and hungry Sudama goes to meet Krishna for assistance. Krishna rushes to embrace him at the door and helps him out of his problems. This impresses upon the minds of children the belief that God looks after those who are devoted to Him and never lets them down. The world and all that goes on in it is *maya*, illusion, and *lila*, or play of the gods.

The ultimate aim of a Bharatanatyam dancer is the realisation of the Universal Being. All Indian art, Vatsyayan (1974) explains, is *sadhana* as means of achieving a state of complete harmony; *yoga* as adeptness or efficiency in the activity undertaken; and *yajna* as offering of the best that one has.

The dance situation provides ample opportunity to introspect and reflect on oneself and those around. Away from the academic programme, crammed with classes and laden with the compulsion to memorise, retain and reproduce in examinations information which appears to be redundant and only of little use, if at all, dance provides an oasis. It fosters co-ordination of the body with the mind, enhancing both receptivity and sensitivity.

This is education for life. It pursues the ideals of an educational system enunciated by Coomaraswamy (1983): (i) universal philosophical attitude; (ii) sacredness of all things, the antithesis of the Western division of life into the sacred and the profane; (iii) religious toleration based on the awareness that all dogmas are formulae imposed upon the Infinite by limitations of the finite human intellect; (iv) etiquette — civilisation conceived of as the product of civil people; (v) relationship between teacher and pupil implied by the terms *guru* and *chela*, in memorizing great literature — the epics — as embodying ideals of character, learning as a privilege never to be used merely as a means to economic prosperity; (vi) altruism and recognition of the unity of all life; and (vii) control not merely of action but also of thought.

People will support the form and content of an educational system founded on their cultural conception, cultural understanding and cultural interpretation. In essence, the concept of education is to be reworked to enfold experience, personal development and self-enrichment as its essential components.

References

Coomaraswamy, A.K., 1983, *Essays in National Idealism*. New Delhi: Munshiram Manoharlal.

Das, S. K., 1986, *The Educational System of the Ancient Hindus*. Delhi: Gian Publishing House.

Saraswati, B.N., 1994, 'Approaching Aboriginal Art', in *The Literary Criterion*, XXIX, 1.

Vatsyayan, K., 1974, *Indian Classical Dance*. New Delhi: Publications Division, Government of India.

———, 1991, 'Literacy and Human Development', in *Human Development: An Indian Perspective*, ed. by K.L. Dalal. New Delhi: Har Anand Publications for UNDP.

11

Buddhism and Education
The Thai Experience

Pataraporn Sirikanchana

SINCE the advent of Buddhism in Thailand nearly 1,000 years ago, monks have had crucial roles especially in the moral education of the public. The Thai mind, in general, is thus inculcated with compassion, friendliness, and love of peace.

Nevertheless, having lacked knowledge of the Buddhist scriptures and modern academic knowledge, the Thais were superstitious and unable to improve their way of life. King Rama V (1868-1910) was aware of his people's inadequate knowledge susceptible to Western imperialism. He began the process of educational reform in the country. The two Buddhist colleges, Mahamakuta and Mahachula, were built in order to properly train monks to be efficient Buddhist teachers and good followers of the *dhamma.* After their graduation, monks went to work in temple schools in villages throughout the country. They taught Buddhist ethics, Thai language, mathematics, history, and so on, and tried to improve folk ways of life. In the reign of King Rama V, Thai people were more educated and contented with their prosperity.

Problems of Education in Thailand Today

Though Thai education has been improved since the reign of King Rama V, the government is still unable to push all children through the process of compulsory formal education. The failure of educational management and administration as well as a rapid increase of population call for non-formal education, e.g., a Temple Pre-school Centre, a Buddhist Sunday School, etc.

It is obvious that only a well-to-do family is able to send children to a good school. Many poor children in Bangkok and those in villages far away are condemned to stay with their parents to work for the rest of their lives. The sight of youngsters selling newspapers and garlands in the street is just an ordinary experience for everyone in Bangkok. Similarly, outside Bangkok, far away in the countryside, small boys and girls watch cattle in the fields without any chance to enter school.

The Roles of the Government and Monks

In order to prepare children for school and provide the socially disadvantaged with the chance to be literate, the Ministry of Education initiated Temple Pre-school Centres in 1963. The project gained good support from the Buddhist Sangha, which allowed any temple to establish a temple school. Monks became teachers of pre-school children. The outcome of the project was successfully accepted by the public.

The Temple Pre-school Centre is a kind of social welfare promoted by the government. It invites people of all ages to come to the temple, be morally cultivated, and fulfil their human qualities. It widely interests the public with its following features:

1. It works on a basis of charity and gives free service.
2. It can be established in any temple.
3. Children eligible are five to eight years old.
4. Any illiterate adult is welcome to enrol.
5. Monks and novices are teachers. Lay people can be teaching assistants.
6. Children are prepared to enter compulsory school and their morals and manners are cultivated.

Since 1988, the Temple Pre-school Centre has changed to 'the Centre of Pre-school Children in the Temple'. It is open to children three-to-six years old.

Apart from the project of the Ministry of Education, there are some other projects for poor children launched by developer monks. One worth mentioning is Phra Khamkhian Suvanno of Sukhato Forest hermitage in north-eastern Thailand.

Phra Khamkhian Suvanno founded the Centre for Child Development in 1978 in order to take care of small children whose parents had to work in the fields all day. Most north-eastern villagers were poor farmers. They had to bring their children to the fields because there was nobody at home to look after them. Waiting for their parents to finish work, children played in the rain or were exposed to the sun the whole day. Some were severely ill and died. Phra Khamkhian thus decided to set up the Centre so that children would be taken care of and would learn to read and write elementary Thai language. At first, there were 20 children in the Centre. Phra Khamkhian brought up these children himself. They were fed with the food given to monks everyday. They had soy milk to drink and sweetmeats to eat regularly. The Centre gave a free service to the community for 8 months of the year from March to November. It was closed on the Buddhist Sabbath days. Some years later, a few volunteer assistants came to teach the children. Parents could leave their children in the centre and go to work happily in the fields.

Apart from 2,554 pre-school centres around the country, the government and the Buddhist order also carry on the project of the Buddhist Sunday School. The Buddhist Sunday School originated in Sri Lanka in 1886. It teaches various fields of Buddhist knowledge and languages. From 1953 to 1957, Phra Bimaladharma of Mahachulalongkorn Buddhist College, Bangkok, had visited Sri Lanka and witnessed moral and cultural teaching in the Sunday School. He deeply appreciated the success of its work. After his return to Thailand, he established the first Buddhist Sunday School in Mahachulalongkorn Buddhist College in 1958.

The Buddhist Sunday School was founded and has been carried on until today in order to inculcate moral discipline and general knowledge in children. Many social problems, e.g., juvenile delinquency, drug addiction, etc., arise from a lack of moral training and moral cultivation. If children are acquainted with the Buddhist teachings and properly follow the Buddhist precepts, they will be able to attain peaceful happiness and live successful lives.

In order to save young people from ignorance and worthless life, Mahachulalongkorn Buddhist College thus began the Buddhist Sunday School with the following objectives:

1. To make young people and children familiar with Buddhism.
2. To inculcate moral discipline and cultural appreciation in young people and children.
3. To teach young people and children to lead their lives according to Buddhist principles.
4. To train young people and children to work for public welfare.

Nowadays there are 995 branches of the Buddhist Sunday School around the country. Class levels are arranged according to students' grades as follows:

1. There are 4 elementary classes for primary school pupils of grades 1, 2, 3 and 4.

2. There are 3 intermediate classes for secondary school students of grades 1, 2 and 3.
3. There are 3 advanced classes for high school students of grades 4, 5 and 6.
4. There are 2 final classes for college students.

It takes 12 years to complete the entire course. Monks and some lay teachers, due to their compassion and loving kindness towards students, work in the programme on a voluntary basis. Their work is much appreciated by all Buddhists and considerably helps improve public morality.

Apart from Mahachulalongkorn Buddhist College's public service of education, there are some other Buddhist leaders, both monks and lay people, who actively work for the sake of disadvantaged children. Worth mentioning here are Phra Vidya Cittadhammo of Mount Sarb Temple School and Mr Pai Soisaklang, the head of Sa Koon village.

According to Thai Buddhist tradition, only a male has the privilege of becoming a novice or a monk and is able to stay in a monastery for further education. A female can be merely a lay attendant and cannot closely associate with monks. Thus, male children have a good chance to fulfil their education through ordination. The case of Mount Sarb Temple School may well illustrate the point.

Phra Vidya Cittadhammo, the developer monk who runs the school, explains that all other secondary schools in the country emphasise only academic knowledge. They pay less attention to the moral cultivation of students' minds. Mount Sarb Temple School, on the contrary, accepts all underprivileged boys whose parents are too poor to send them to a formal school. These children are ordained in the temple and have studied in its secondary school, free of charge, for 3 years. The subjects are Thai, English, mathematics, and social studies. Children and young people who are novices hold to the ten Buddhist precepts during their 3 years of education, which covers both secular knowledge and *dhammic* knowledge.

It is very important nowadays to pay attention to juvenile development. Samanera Vidya, a young novice, points out that due to his poverty, he cannot enter any other school and that most youngsters are wayward and delinquent. He stresses that the Mount Sarb Temple School tremendously helps young people cultivate their minds, fulfil their human qualities, and be able to survive happily in this suffering world.

Another example of service for young students is the Sa Koon Village School, founded by the headman Pai Soisaklang and all villagers. Pai Soisaklang uses Buddhist teachings as a guide for village life. He and the villagers built the school, without any support from the government, in order to prevent youngsters from wrongdoing. By means of education, children learn to differentiate right from wrong and be sufficiently knowledgeable to depend on themselves. He also encourages the villagers to live according to the Buddhist precepts, for their own peaceful happiness, through the following village rules:

1. Do not kill animals, since it is forbidden by the first Buddhist precept.
2. Do not fire a gun within the village since it will frighten others.
3. Do not drink alcoholic beverages in the village since the drunkenness may cause trouble to others.

During the Buddhist Lent, young people are persuaded to listen to a sermon and join a religious ceremony in the temple. They learn to stay close to their parents and follow the traditional way of life. During weekends, children stay home to help their parents work in the fields and dig fish-ponds. They are happy to make themselves useful to their community.

A Look to the Future

The changing role of Thai monks from ascetics to social developers is indispensable for Thai society nowadays. This phenomenon does not diminish monks' sacred status at all since they still preserve their monastic discipline.

Once when I spent a week at the Sukhato Forest Hermitage in north-eastern Thailand, I had the opportunity to examine Phra Khamkhian Suvanno's community development. I found that his work was much beneficial to the villagers and indispensable for community life. The three villages around the hermitage are located quite far away from the helping hand of the government and cannot survive without monks' assistance.

Phra Khamkhian's method of work is a combination of *dhammic* practice and social development. He follows the Buddhist teaching that a good mind yields a good practice. If one has learned to purify oneself and is able to lessen one's own defilements, one can live for the sake of others and thus can yield benefits to one's community and to the rest of the world. Phra Khamkhian teaches the villagers to abstain from all evil deeds and to practise meditation in order to learn more about themselves and to understand the nature of the world.

Under the sponsorship of the Thai Khadi Research Institute, Thammasat University, Bangkok, I spent nearly 2 months interviewing many villagers, monks, and government officials who were responsible for the well-being of these villages. The outcome of Phra Khamkhian's work, however, is unsatisfactory. Most projects are initiated and run only by

Phra Khamkhian. Villagers are merely participants and thus have no motivation to carry on and fulfil their work. For example, even though the Centre for Child Development still operates today, Phra Khamkhian is the only one who manages it and takes responsibility for all the work. Due to poverty and the lack of self-dependent orientation, the villagers would rather leave all problems to Phra Khamkhian than participate in the Centre or donate some money to support it.

Phra Khamkhian is well conscious of his role as a Thai monk venerated by lay people. His community development has been done within the context of Buddhist discipline. Thus, his monastic status is always held as sacred by the villagers.

Furthermore, it can be noted that most private agencies working for community development do not like to communicate or seek assistance from governmental officials. They prefer working by themselves. The case of Phra Khamkhian is an example of a leading developer monk who devotedly works for the sake of the poor community but lacks internal cooperation and external support.

In Thailand the traditional system of primary education is as important as the modern system since ethical training is no less crucial than academic proficiency. The Thai traditional system of primary education began in a temple or a monastery where monks were teachers and preachers. Its aim was to moralise the public as well as to improve the folk ways of life. According to Buddhist beliefs, the cultivation and purification of the mind is the source of all good deeds. Thus, if we properly bring up children at the earliest, they will become good citizens and good human beings in the future. Their knowledge will be applied to save the world. Right now, the Ministry of Education in Thailand is well conscious of this fact and mandates that all primary school pupils should study Buddhist ethics and should be trained to be morally good in their own traditions. It seems that the traditional system of education emphasises the role of religion for the good of students while the modern system honours academic knowledge and Westernization as signs of educational achievement.

I believe that both moral cultivation and academic training are equally important for students. We would rather have a morally knowledgeable person than an evil brilliant guy or a virtuous idiot. In order to train children both morally and academically, the government and all private agencies need to cooperate in supporting educational projects in all schools and in providing all illiterates throughout the country with compulsory education and elementary knowledge suitable for their folk lives and local environment.

An ideal primary school or centre of education should be well equipped with pictorial lessons and audio-visual materials in order to draw pupils' attention and make the entire process of teaching attractive. The size of the centre depends on the number of pupils in a community. Nevertheless, it is better to have 15-20 pupils in each class so that the teacher is able to train and look after everyone. Children who are ready to enter a primary school and be able to fulfil their course are ideally 5 years old. The curriculum should emphasise knowledge useful for children, encouraging optimistic viewpoints and leading to the fulfilment of their potential. The characteristics of a teacher and parents' education are no less important. Teachers should love children and be cheerful enough to make the lessons interesting. They should be trained particularly for their profession. Needless to say, if the government really wants to support public education and fight against illiteracy, it needs to provide the public with free and compulsory primary education. Since the children of today are the adults of tomorrow, as a Thai proverb says, the attempt by all means to make the best of children is thus worth the investment.

12

Education For Life and Through Life

Gandhi's *Nayee Talim*

Devi Prasad

Education in India under Colonial Rule

As part of her struggle for freedom from colonial rule, India needed a fresh approach to educational planning. It was necessary, first of all, to get rid of the educational system the colonial rulers had imposed on the country after destroying its indigenous traditions and institutions, and second, to design its own educational system for helping to build a genuinely liberated and egalitarian India.

By the middle of the nineteenth century a significant degree of awareness of their own identity had developed among the educated elite of the country. It was the dawn of the period of social reformation and growth of Indian nationalism. There were two distinct ways of looking at the issues of reconstructing the social fabric of the country at that point in time of the freedom struggle. Some of the leaders were of the opinion that to be able to rise to the level of the British rulers Indians must educate themselves in the English system. They established educational institutions which aimed at the particular goal. Although they too were interested in discovering their original roots, they probably felt that the British were able to rule India because of their superiority over us in many ways.

On another level movements grew to motivate people to go back to their Vedic culture, in other words, to discover the classical roots of Indian civilisation. They founded *gurukuls* and other traditional institutions of learning and teaching. According to them India had become weak because she had drifted away from her classical way of life. Neither of these approaches faced the real issues, issues that were related to the question of real liberation and self-reliance.

India had been severely impoverished and her population demoralised. The Indian masses had lost the taste for freedom almost completely. Until the end of the first half of the nineteenth century, the military might of the whole country, if it could have been put together, was considerable. But during the period after the 1857 soldiers' mutiny, the morale of the people had fallen to a very low ebb. Moreover, the British rulers, having realised the tenacity of the Indian soldiers and realising that militarily they could not enslave India permanently, cunningly and systematically disarmed the nation. So much so that the simple farmer could not own an ordinary gun to protect his crops from the menace of wild and stray animals.

Industrially India, according to Romesh Dutt, was one of the leading countries of the world until the middle of the eighteenth century. But by the end of the nineteenth century Indian industry had been nearly totally destroyed. The country which supplied a great variety of textiles to many countries of the world, specially northern Europe, was now importing cloth from Britain for most of its needs.

Several nations used to get their ships built by Indian ship-builders. Richard Greg quotes in *White Sahebas in India* the factor of the East Indian Company, who wrote in 1670:

> Many English merchants, and others have their ships and vessels yearly built. Here is the best and well grown timber . . . best iron upon the coast. . . . They have an excellent way of making shrouds, stays and any other rigging for the ships.

Lord Wellesley wrote in 1800:

> The port of Calcutta contains about 10,000 tons of shipping, built in India of a description calculated for the conveyance of cargo to England.

Taylor, in his *History of India*, quoted from Mrs Besant's *India Bond or Free*:

> the arrival in the Port of London of Indian produce in Indian-built ships created a sensation among the monopolists which could not have been exceeded if a hostile fleet had appeared in the Thames. The ship-builders of the Port of London took the lead in raising the cry of alarm . . . . An obliging Government saw to it that the Indian industry perished.

The creation of the worst kind of landlordism damaged human relationships and divided society into strata which were completely alien to the Indian experience. The planned destruction of the *panchayat*, the system of local government and administration, including the judiciary, had even a worse effect on social relationships.

The colonial rulers could not have been satisfied only by destroying India's industry, agriculture, administrative and legal institutions and their national defence system. They were hoping to enslave India permanently. The British knew that only material enslavement was not sufficient. Psychological enslavement was the most important and effective mechanism to reach that goal. Thus education became the biggest prey for the Raj.

The colonisers took immediate steps to build a long-term strategy based on 'educational reforms'. They took every possible step to weaken the educational traditions which were deeply ingrained in the culture of the people and were the most powerful instruments that had retained cultural values among the masses for such a long period.

We should also note the attitude of the colonisers to the literature of Asian countries. In his famous minutes Lord Macaulay wrote:

> I have no knowledge either of Sanskrit or Arabic, but I have done what I could to form a correct estimate of their value. I have read translations of the more celebrated Arabic and Sanskrit works. I have conversed both here and at home with men distinguished by their proficiency in the Eastern tongues. I am quite ready to take Oriental learning at the valuation of the orientalists themselves. I have never found one among them who could deny that a single shelf of a good European library was worth the whole native literature of India and Arabia.

He added:

> We must at present do our best to form a class who may be interpreters between us and the millions whom we govern — a class of persons Indian in blood and colour, but English in tastes, in opinions, in morals and in intellect.

The Governor General, Lord Bentinck, said:

> I give my entire concurrence to the sentiments expressed in minutes.

In spite of such a calculated onslaught on Indian culture, specially its educational traditions, a significant proportion of institutions had survived until at least 1835. Adams, who was an officer in the Education Department of Bengal, conducted an inquiry into the number of schools in that region at that time. According to his calculations there was a school for every thirty-two boys and these schools were provided to most of the 150,000 villages. Writing about the Indian educational system of that time, Max-Müller said:

> There is such a thing as social education and education outside the books; and this education is distinctly higher in India than in any part of Christendom. It is an education not in the so-called three R's, but in humanity.

There are documents prepared by some British officials of that period to show that in several regions of the country literacy was very high — 100 per cent among the male population and not very much lower among females. It is the same country in which literacy had fallen to less than 10 per cent by the end of the last century. By and large, India then was considered an illiterate nation!

National Awakening

The end of the last century saw the re-emergence of movements for national awakening, but this time of a different kind — very different from the wars of Tipu Sultan and the Sepoy Mutiny. The emergence of the Arya Samaj, the Prarthana Samaj and the Brahmo Samaj brought about much awareness among the people of their cultural heritage, which gave them a sense of self-identity.

The Indian National Congress, which had a political character, was formed in 1885. Although it started talking of Indian self-rule, Indians' participation in the administrative services, etc., the majority of its leadership was English educated and did not speak the language of the people, who did not understand what the Congress leadership was talking about. As these leaders had no contact with the masses, the Congress could not command a popular following.

During the beginning of the growth of the Indian National Congress the British took advantage of the situation and, to some extent, were able to co-opt some of its leadership to neutralise their effect, if any, on the masses. But this time they could not stop the tide of liberation. They could only slow it down by their classical technique of divide and rule.

Gandhiji's Educational Work in South Africa

Before going into Gandhiji's educational experiments in India we should briefly discuss his educational experiences in South Africa.

While mobilising the Indians living in South Africa for the struggle for self-dignity, Gandhiji had gone through various experiences and had conducted experiments to find an alternative approach to conflict resolution and re-structuring social relationships through non-violence. Among these experiments those on education are the most important for us here.

After reading *Unto This Last* he said:

> I believe that I discovered some of my deepest convictions reflected in this great book of Ruskin, that is why it so captured me and made me transform my life.

The teachings of the book as he grasped them were: (a) that the good of the individual is contained in the good of all; (b) that a lawyer's work has the same value as the barber's, inasmuch as all have the same right of earning their livelihood from their work; (c) that a life of labour, i.e., the life of the tiller of the soil and the handicraftsman, is the life worth living. This experience put him on to a new track more intensely and with greater commitment to search for a better life.

Gandhiji was also editing and publishing *Indian Opinion* from Durban, which he thought should be moved to a farm, on which everyone should labour, drawing the same living wage, and attending to press work in their spare time. After a discussion with his colleagues about his ideas, with which they all agreed, *Indian Opinion* was moved to Phoenix, fourteen miles from Durban. Thus the Phoenix Settlement was started in 1904. It became a well-knit family of committed people trying to live their lives under the guidance of Gandhiji.

He was very conscious about the need for the education of the members of the Phoenix Farm, specially of children. He wrote in his autobiography:

> As the Farm grew, it was found necessary to make some provision for the education of its boys and girls. There were among these, Hindu, Musalman, Parsi and Christian boys and some Hindu girls. It was not possible, and I did not think necessary, to engage a special teacher for them. It was not possible, for qualified Indian teachers were scarce, and even when available, none would be ready to go to a place 21 miles distant from Johannesburg on a small salary . . . .I did not believe in the existing system of education, and I had a mind to [find] by experience and experiment the true system.

Gandhiji wrote in his autobiography:

> When I landed at Durban in January 1897, I had three children with me. My sister's son aged 10, and my own sons aged 9 and 6. Where was I to educate them? I could have sent them to the school for European children, but only as a matter of favour and exception. For no other Indian children would be allowed. For these there were schools established by Christian

> missions, but I was not prepared to send my children there, as I did not like the education imparted in those schools. For one thing, the medium of instruction there would be only English, or perhaps incorrect Tamil or Hindi, which too could be arranged not without difficulty. I could not possibly put up with this and other disadvantages. I was making my own attempt to teach them but that was at best irregular and I could not get hold of a suitable Gujrati teacher . . . . I was at my wits' end. . . .

Gandhiji's struggle continued. On the one hand he totally rejected the existing system of education, but on the other he did not have a clear alternative to replace it. He had no idea of how to go about the task of educating the children of his extended family. However, he was deeply convinced that only that education is desirable which develops a healthy self-image and inculcates certain values in the individual. He wrote:

> Had I been blind to a sense of self-respect, and allowed myself to be satisfied by having for my children the education that other children could not get, I should have deprived them of the object-lesson in liberty and self-respect that I gave them at the cost of literary training. And where a choice has to be made between liberty and learning, who will not say that the former has to be preferred a thousand times to the latter.

After getting the journal well settled he wanted to start a school for Indian children. On 13 January 1905 he wrote a letter to Professor Gokhale asking for his support. Following are a few sentences from it:

> It is also my intention, if my earnings continue, to open a school on the grounds, which would be second to none in South Africa, for the education primarily of Indian children who would be resident boarders and, secondly, of all who want to join the school but would also reside on the premises. For this too volunteer workers are required. It would be possible to induce one or two English men or English ladies here to give their lifetime to this work, but Indian teachers are absolutely necessary. Could you induce any graduates who have an aptitude for teaching, who bear a blameless character and who would be prepared to work for a mere living. Those who would come must be well-tried first-class men. I would want two or three at least but more could certainly be accommodated, and after the school is in working order, it is intended to add a sanatorium with open-air treatment on hygienic lines. . . .

The school eventually had fifty children. He was determined 'to find out by experience and experiment the true system'. He took some time to discover it, specially the kind that could be offered for wider application in a country so large as India, for he knew that his home was India and not South Africa.

Gandhiji Returns to India

Gandhiji eventually realised that it was time for him to return to his home country. He reached Bombay with Kasturba on 9 January 1915.

Wise man that he was, Gandhiji decided not to straight away plunge into the politics of the country or even pass any judgement on the situation here. He gave a year to going around the country to see and acquaint himself with the life of the people and their feelings. He wanted to have first-hand knowledge of the rural as well as urban conditions in which the common man and woman lived. This was also the advice given to him by Gopal Krishna Gokhale, who was like a political *guru* to him. The question of discovering the right kind of educational system, however, remained his primary concern.

Within two or three years Gandhiji had to undertake three major *satyagraha* campaigns — the Ahmedabad Satyagraha related to the textile mill-hands' strike, the Kheda Satyagraha about the payment of land revenue by farmers, and the Champaran Satyagraha against the exploitation of the indigo farmers by the British plantation owners. The Champaran Satyagraha was Gandhiji's first major non-violent struggle in India. It was aimed at uplifting the physical conditions and morale of the indigo farmers and liberating them from the grip of the British indigo plantation owners who were exploiting them in a most inhuman manner. This *satyagraha* proved to be the first profound victory of non-violent struggle against the exploitation of the helpless by the white plantation owners. However, when the *satyagraha* was still going on, he was thinking of improving the conditions of the poor of Champaran district.

In *Satyagraha in Champaran* Rajendra Prasad wrote:

> He [Gandhiji] was convinced from the very beginning that it was impossible for any outside agency to improve their lot unless their mental and moral condition was improved. This applies to the whole of India but it can be demonstrated beyond contradiction in Champaran . . . . Mahatmaji had accordingly decided that arrangement for spread of education was as necessary among them as the redress of their grievances. Some time before the Enquiry Committee commenced its work Mahatmaji had written to

> some friends about it and told them what sorts of volunteers he needed for social work.

He did get some of the most educated people from various parts of India to run the schools which he had managed to start. Unfortunately very few educated people from Bihar joined him. Describing the plan he wrote to a Government official:

> In the schools I am opening, children under the age of 12 only are admitted. The idea is to get hold of as many children as possible and to give them an all round education, i.e. a good knowledge of Hindi or Urdu, [and] through that medium, of arithmetic and the rudiments of history and geography, a knowledge of simple scientific principles and some industrial training. No cut and dried syllabus has yet been prepared because I am going on an unbeaten track. I look upon our present system with horror and distrust. Instead of developing the moral and mental faculties of the little children it dwarfs them. . . .

Five schools started functioning in the area. Some of his closest companions were with him to carry out the plans, about which he wrote:

> I shall endeavour to avoid the defects of the present system. The chief thing aimed at is contact of children with men and women of culture and unimpeachable moral character. That to me is education. Literary training is to be useful merely as a means to that end. . . .

The Champaran Satyagraha came to a successful end. In his autobiography Gandhiji wrote:

> The *ryots* [tenants], who had all along remained crushed, now somewhat came to their own, and the superstition that the stain of indigo could never be washed out was exploded . . . . It was my desire to continue the constructive work for some years, to establish more schools and to penetrate the villages more effectively. The ground had been prepared, but it did not please God, as often before, to allow my plans to be fulfilled. Fate decided otherwise and drove me to take up work elsewhere.

Those schools functioned for a while but were then closed for lack of workers and the vision required for such a pioneering work.

Satyagraha Ashram, Sabarmati

When leaving South Africa Gandhiji had an important question in his mind. Where should the Phoenix family settle in India? At the suggestion of C.F. Andrews the whole group went to Santiniketan, the *ashram* poet Rabindranath Tagore had set up in Bengal. Tagore had a comprehensive educational programme, which, to put it briefly, had a precise philosophy and was built around creative activities. He was firm in his conviction that the medium of instruction at all levels should be the mother tongue. Second, for him Nature was the richest centre of learning; and third, creative activities ought to play a central role in the processes of education. Gandhiji admired and respected Tagore's educational system, but it has to be admitted that his goal was somewhat different, particularly his short-term goal.

The Phoenix 'settlement' shifted to Santiniketan. After a few days Gandhiji also joined them. He tried various experiments with the help of the poet and the teachers, specially about the way of life educational institutions should develop. No doubt he must have drawn some creative ideas from his and his colleagues' experience in Santiniketan. There is not much left of those experiments except that every year a day is observed to recall Gandhiji's stay and the programme of self-help.

Santiniketan was not the place for Gandhiji to make a home and conduct his own experiments. His need was to experiment with the perspective he had developed for the future of India. The 'family' went to Gujarat, first to Kochrab and finally to Sabarmati, a place just outside Ahmedabad.

In addition to his preoccupation with national politics there was one subject which occupied his mind all the time. It was education for the masses of India. He knew what 'education' meant in pre-British India. At a conference in Allahabad on 23 December 1916 he gave a talk which was reported in *The Leader* (27.12.1916). Following are a few sentences from that report:

> Mr Gandhi then described the ancient system of education . . . even elementary education imparted by the village teacher taught the students all that was necessary for their occupation. Those who went in for higher education became fully conversant with the science of wealth, *Artha Shastra*, ethics and religion, *Dharma Shastra*. In ancient times, there were no restrictions on education . . . . It was due to such a system of education that Indian civilization had outlived so many vicissitudes through thousands of years . . . . No doubt the wave of a new civilization had been passing through India. But he was sure that it was transitory, it would soon pass away and Indian civilization would be revivified.

Gandhiji was steadily getting closer to his ideal educational system. But he had yet to try many more experiments before he discovered **the true system**.

National Education

After the Sabarmati Ashram was well settled, Gandhiji began planning for a national school. Describing its basic principles in the prospectus he wrote:

> The education will be physical, intellectual and religious. For physical education there will be training in agriculture and hand-weaving and in the use of carpenter's and blacksmith's tool. . . . In addition, they will be given drill, . . . and as part of this, they will be taught how to march in squads and how each one may work with quiet efficiency in case of accidents such as fire. . . . They will have instructions on how to preserve health and on home remedies for ordinary ailments, with as much of physiology and botany as may be necessary for the purpose . . . . For intellectual training, they will study Gujarati, Marathi, Hindi and Sanskrit as compulsory subjects . . . . There will be no teaching of English during the first three years".

The medium of instruction was to be Gujarati (the mother tongue and local language) right up to the highest stage. The syllabus included the usual subjects such as mathematics, book-keeping, history, geography, chemistry and astronomy. By way of instruction in religion, general ethical principles were to be taught. The aim of that education was that after a few years the student's equipment would approximate to that of a well-informed graduate.

It is obvious that although he was more or less clear about the objectives of his ideal education, he had not yet reached the stage when he could put forward a concrete plan of his system which, by definition, had to be entirely different from the existing one with the structure the British had created.

He was firm about education being imparted through the mother tongue at all its stages. In a speech at the Second Gujarat Educational Conference in October, 1917, he said:

> It should be obvious to everyone that the first thing to do in this connection is to come to a definite decision about the medium of instruction. Unless that is done, all other efforts, I fear, are likely to prove fruitless.

He put equally great stress on the desirability of having a common language for the whole country. He enumerated five requirements of a national language. It should be: (a) easy to learn for government officials; (b) capable of serving as a medium of religious, economic and political contact between all parts of the country; (c) the speech of a large number of Indians; (d) easy for everyone to learn; and (e) chosen without consideration of temporary or passing circumstances. It was evident that English did not fulfil any of these requirements. Gandhiji considered Hindi, eventually Hindustani, to be the language that fulfilled all the requirements.

Gandhiji emphasised the building of character as the major task of education. The third item on which he put emphasis was music and the fourth was physical training. He strongly pleaded for the education of women.

We think our entire life depends on success at examinations. Gandhiji said:

> India never knew the institution of examination. The method is of recent introduction . . . . The system has lent itself to serious abuse, every subject being taught with an eye on the examination and the conviction firmly planted in the pupil's mind that passing the examination was all that was necessary.

In his scheme it was stated:

> Having regard to the view that examinations are quite undesirable, pupils in this institution will be tested periodically from two points of view — whether the teacher has made the right effort and whether the pupil has followed. The pupil will be freed from the fear of examinations. . . .

National Universities — Vidyapeeths

After the First World War India waited for a change of heart on the part of the British. Instead she received the Rowlatt Act. The people as a whole resented it. In April 1919 more than a thousand innocent unarmed people were massacred in Jallianwalla Bagh, Amritsar, by the army. This outrage wounded India's pride and she rose in revolt against foreign domination. Gandhiji said that Jallianwalla Bagh was only the beginning:

> We must be prepared to contemplate with equanimity not a thousand murders of innocent men and women, but many thousands before we attain a status in the world that shall not be surpassed by any nation. We hope, therefore, that all concerned will take rather than lose heart and treat hanging as an ordinary affair of life.

On 1 August 1920 he wrote to the Viceroy: "I can retain neither respect nor affection for a government which has been moving from wrong to wrong in order to defend its immorality". The Congress and Gandhiji declared total non-cooperation with the British Government. The call included a total boycott of legislative bodies, Government schools and law courts.

The Gujarat Political Conference held on 27-29 August passed a resolution which meant that title-holders and persons holding medals should renounce the titles and return the medals; lawyers should try to settle disputes privately and should give up practice in law courts; parents should withdraw their children from any school that had any connection with the Government; college students should give up colleges on their own; voters should refuse to vote for any one seeking election to a council and candidates should withdraw their candidature; and every man, woman and child should follow the rule of perfect *swadeshi* and should spin cotton.

Although it was a solely political campaign, it had a profound impact on the educational scene in the country. With the call for boycotting schools and colleges having any connection with the government, it was necessary to have some alternative institutions. That was the beginning of the establishment of *vidyapeeths* — national schools and colleges all over the country. In November Gandhiji founded the National University of Gujarat, Gujarat Vidyapeeth. Such national institutions also came up in Calcutta, Patna, Aligarh, Bombay, Banaras and Delhi. The Jamia Millia Islamia — the National Muslim University — was founded through the joint efforts of Gandhiji and Md. Ali.

These institutions made a significant contribution in the fields of both political awakening and educational awareness. Establishing these centres of learning was a step towards the growth of a new approach to education. They created the spirit of freedom, an element essential for the development of a healthy personality. Second, they prepared a cadre of freedom-fighters for the next 'battle for liberation'. More important, as experimental centres they were useful for Gandhiji to explore the structural side of his educational scheme, *buniyadi talim* (basic education), which was started in 1937. By that time institutions like Gujarat Vidyapeeth and Jamia Millia Islamia had fully developed their systemic structures in a way that was favourable for giving a trial to the new scheme.

Nayee Talim

Between 1920 and 1937 a lot happened in the country. People's awareness of their cultural foundation, their knowledge of the damage done to it by the colonial rulers, and an understanding of the socio-political conditions of the country made it imperative that for building a healthy India mass education in general and education of the individual along with the struggle for freedom were the most important elements. Gandhiji also believed that without gaining freedom, building a new and independent educational programme was not possible.

Most of the revivalist institutions, barring some exceptions, such as Santiniketan and *vidyapeeths*, had lost their original relevance or had become dependent on government recognition. Some continued catering for the needs of the elite, who looked for government patronage. Santiniketan was deliberately kept apolitical by Tagore, except that some individuals occasionally took active part in the freedom struggle. Without going into a discussion on the why of these issues I shall try to see the relationship between the Tagorean concept of education and *nayee talim* as developed under the guidance of Gandhiji.

As mentioned earlier, according to Tagore's scheme there are three centres of education: the mother tongue, not only as the medium of instruction but also as the major means of communication between people; creative activities; and Nature, of which we are an inseparable part.

Gandhiji's scheme as presented to the nation in 1937 also had three similar elements. He was convinced that the medium of instruction in schools and colleges must be the mother tongue or the language of the area. He also insisted that the language of communication between people of various regions ought to be that which most people understood or could easily learn. In other words, it should be an effective tool for community relationships. For most of its needs, physical as well as spiritual, human society depended on Nature. *Nayee talim*, as envisaged by Gandhiji, gave top priority to our relationship with Nature. According to Tagore Nature was the most important source of knowledge, human creativity and livelihood. The highest education is that which does not merely impart information but puts life in harmony with all existence.

The third element is meaningful manual work according to Gandhiji, and creative activities according to Tagore. There is very little difference in the two approaches. Tagore did not talk of the economic side of creative activities. For him it was art, something that caters for all human needs, physical, emotional and spiritual. Gandhiji, on the other hand, placed emphasis on it as a vocation for livelihood and a way of gaining knowledge.

I do not think the difference between the two is crucial. While Gandhiji's approach is obviously egalitarian, Tagore's sounds elitist. Taking into account the time factor, Gandhiji had the advantage of drawing from Gurudev's educational experience. He did not only learn something but he also improved on it.

Basic Education Starts in Seven Provinces

The fruit which Gandhiji had nurtured for two decades was now nearly ripe. The political situation in India had changed. After the elections for Legislative Assemblies the British Government had to accept the formation of Congress ministries in nine provinces. In addition to whatsoever these Congress Governments were hoping to do, Gandhiji offered them an educational plan to replace the current system of education. He had written an article, 'Education', in the 31-7-1937 issue of *Harijan*.

> How to solve the problem of education is the problem unfortunately mixed

> up with the disappearance of the drink revenue. . . .

Until then the expenditure on education came from excise (alcohol) revenue. Gandhiji wrote:

> . . . as a nation we are so backward in education that we cannot hope to fulfil our obligation to the nation in this respect in a given time during this generation, if the programme is to depend on money. I have therefore made bold, even at the risk of losing a reputation for constructive ability, to suggest that education should be self-supporting. By education I mean all-round drawing out of the best in child and man — body, mind and spirit . . . .

He suggested that the money required for education could be raised by taxing the rich.

This is not a fanciful picture. If we would but shed our mental laziness, it would appear to be an eminently reasonable and practical solution of the problem of education that faces the Congress Ministers and therefore the Congress.

In October 1937 Gandhiji called a conference which was attended by many eminent education experts of the country and the Education Ministers of several provinces. He presented his scheme to them. The conference passed a resolution which agreed that free and compulsory education should be provided to every child of seven to fourteen years of age; the medium of instruction should be the mother tongue; the process of education throughout this period should centre around some form of manual productive work, and all the other abilities to be developed should be integrally related to the central handicraft chosen, with due regard to the environment of the child; and this system of education would gradually be able to cover the remuneration of the teachers.

A little later the Hindustani Talimi Sangh, an all-India organisation, was formed to develop the Basic Education programme all over the country and run experimental schools. The first such school was set up in Sevagram, Gandhiji's *ashram*, which also became the central office of the Sangh with Shri E.W. Aryanayakam as General Secretary. Gandhiji chose Dr Zakir Hussain, the head of Jamia Millia Islamia, as the Chairman of the Sangh.

A wave of educational reconstruction seemed to pass over the country. Some provinces appointed education reorganization committees, teacher training and refresher training centres. Some basic schools were opened and some primary schools were converted into basic schools. New literature on Basic Education was published for the use of teachers. A seven-year syllabus was prepared by some of the most experienced teachers of their subjects.

In his report of two years of work Aryanayakam wrote that

> basic education was being carried out as an educational experiment by the Governments of C.P., U.P., Bihar, Orissa, Bombay, and Kashmir and a few non-Government institutions. In all there were twelve training schools and two training colleges, seven Refreshers Training Centres and over five thousand schools carrying out the experiment of basic education.

There was so much enthusiasm that even when the Orissa Government closed its basic schools in the second year of the experiment, the people of the province continued the work on their own. Out of the fifteen basic schools, seven continued functioning, which indicates that the system had attracted the interest of the people.

After about two years the Congress Ministries resigned, resulting in the closure of basic schools run by the governments in all the seven provinces. But the ones run by voluntary bodies continued until the launch of the Quit India struggle in 1942, the consequence of which was the arrest of most of the active workers, including those engaged in Gandhiji's constructive programmes. He was the first to be arrested.

The resolution passed by the Second Basic Education Conference held on the campus of Jamia Millia Islamia gives an idea of the results attained by the basic schools during the short period of a little over two years.

> This conference records with satisfaction that the reports on the working of basic schools run by the Governments, local bodies and by private enterprise are almost unanimous that general standards of health or behaviour as well as intellectual attainment, are very encouraging. The children in basic schools are more active, cheerful, self-reliant, their power of self-expression is well-developed, they are developing habits of cooperative work, and social prejudices are breaking down. Considering the difficulties inherent in the initial stages of a new scheme of education, involving a new ideology and a new technique, the progress reported holds out the promise that even better results can be expected in future.

Nayee Talim is Born

The Quit India movement of 1942 proved a very intense experience for the country, specially in regard to looking back and reviewing the achievements and difficulties faced by the freedom struggle led by Gandhiji. Within a very few days most of the cadres of the freedom movement had been arrested. Only those remained outside who either went underground or were co-opted — willingly or otherwise — by the State machinery. Members of political parties which were not in favour of the Quit India Call also remained out of prisons.

Gandhiji was in prison for nearly two years. That was probably the hardest period in his life. His two closest life companions died in the same prison — his wife Kasturba and Mahadev Desai, his right hand. During that period he must have introspected and reviewed almost everything he did in his life or wanted to do. His life-long *tapasya* as a teacher must have made him ask why he had not yet been able to find **the right way**, which he had been seeking since he started teaching the children of his large family in South Africa. He must have asked himself many questions including some regarding his experiment of basic education and its future.

When he came out of prison, two of his major concerns were (a) to see the British rule over India end at the earliest, and (b) to plan an educational programme for the people of India which would prepare them for using that freedom to lift themselves up from the conditions they had been pushed into during colonial rule, such as stark poverty, pessimism, ignorance and helplessness.

For the last quarter of a century Gandhiji had led the non-violent war of freedom on political as well as socio-economic grounds. He went on encouraging the people to build alternatives for nearly all aspects of life. The eighteen-point Constructive Programme was for replacing the existing systems and institutions. For instance, development of indigenous industries — textiles, housing, food, basic education, health and sanitation, social equality and cooperative living.

Whenever a suggestion came to him for improving or changing some aspect of Indian life, if he felt that the suggestion warranted attention, he found people, or people came to him to take up the responsibility to organise that particular activity. The building of the leprosy campaign is a remarkable example of that kind. Traditionally leprosy was a most misunderstood disease. Someone with that disease was considered lower than an untouchable. Gandhiji had always been working against such traditions on both medical and socio-economic grounds. In 1946 a man who had the disease and who had cured himself went and told Gandhiji his story. I remember having heard him say to this gentleman: "You have been sent to me by God with the plan that you will organise and lead an all-India movement against this most inhuman attitude towards leprosy". It inspired the man so much that he built an all-India leprosy campaign. Thus it became one of the eighteen items of the Constructive Programme.

Gandhiji's Constructive Programme was built to provide the country with alternative techniques and institutions to replace the ones imposed by the colonial rulers, and with which the country did not wish to continue after attaining freedom. He had hoped that once India became free she would have a fully tested pattern of political, economic and social structures to run the country on a sound non-violent basis.

His internment in the Aga Khan Palace, as has been said before, was a time of introspection for Gandhiji. After coming out of prison he said: "I have been thinking hard during the detention over the possibilities of Nayee Talim until my mind became restive". He also said:

> We must not rest content with our present achievements. We must penetrate the homes of the children. We must educate their parents. Basic education must become literally education for life. . . . It had become clear to me that the scope of basic education has to be extended. . . . A basic school teacher must consider himself a universal teacher. His village is his universe . . . .

A *nayee talim* conference was called in the month of January 1945. In his inaugural speech Gandhiji introduced an entirely revised and enlarged map of the system. Addressing the basic education workers he said:

> Although we have been working for Nayee Talim all these years, we have so far been, as it were, sailing in an inland sea which is comparatively safe. We are now leaving the shoals and heading for the open sea. So far our course was mapped out. We have now before us uncharted waters with the Pole Star as our only guide and protection. That Pole Star is village handicrafts.
>
> Our sphere of work now is not confined to Nayee Talim of children from seven and fourteen years; it is to cover the whole of life from the moment of conception to the moment of death. This means that our work has increased tremendously. Yet workers remain the same. But that should not worry us. Our guide and companion is Truth which is God. He will never betray us. But Truth will be our help only if we stand by it regardless of everything. There can be in it no room for hypocrisy, camouflage, pride, attachment or anger.
>
> We have to become teachers of villagers; that is to say, we have to become their servants in the true sense. Our reward, if any, has to come from within and not from without. It should make no difference to us whether in our quest for Truth we have any human company or not. Nor does Nayee Talim depend on outside financial help. It must proceed on its own way, whatever critics might say. I know that true education must be self-supporting. There is nothing to feel ashamed of in this. It may be a novel idea if we can make good our claim and demonstrate that ours is the only method for the true

> development of the mind. Those who scoff at Nayee Talim today will become our ardent admirers in the end and Nayee Talim will find universal acceptance.
>
> Whether this is a mere dream or a practical reality, this is the goal of Nayee Talim and nothing short of it. May the God of Truth help us to realise it.
>
> I want to draw your attention also to another thing. I consider the Sevagram centre to be an ideal centre for conducting the central experiment in Nayee Talim, as it is here that the Charkha Sangh (All Indian Spinners' Association) is carrying out its main experiment. Wardha is the centre for the other village industries .... Sevagram does not stand alone; there are nearly 20 villages lying about it in close proximity. Therefore if experiments in Nayee Talim in its most natural form can be carried out anywhere, it is here.

The plan he put forward was for the education of everybody in the community. It was divided as follows: (i) Adult education — of the whole community, including the parents of unborn babies; (ii) pre-basic — education of children between two and a half to seven years; (iii) basic education — for children between seven and fourteen years; (iv) post-basic education — fourteen to eighteen years; (v) university level education — beyond eighteen years of age.

The most significant statement from Gandhiji at that time was that after the launching of the new perspective of *nayee talim* every constructive programme must have its orientation. To put it the other way round, it would mean that all those activities should be considered a part of *nayee talim.* In other words, all the work that was being done for the attainment of freedom for the country and for the reconstruction of the Indian polity should be educationally oriented.

Nayee Talim at Sevagram

Several organisations which had been engaged in basic education earlier restarted their work. The main centre in this regard was Sevagram. I was fortunate to have been offered a position as art instructor for the first six-month teachers' training camp with a very active basic school as the practising school for the teacher trainees and a model school for the country wide *nayee talim* work. It feels strange that instead of six months I continued with *nayee talim* at Sevagram for eighteen years! When I reached Sevagram in November 1944 I remember noticing the enthusiasm of the teachers and the lively faces of the children. The Teachers' Training Camp had more that sixty trainees from almost every part of the country. It was indeed a bright and active group of people of ages between twenty and fifty.

After the prayer at six in the morning and then breakfast, we divided ourselves into groups to do the cleaning of every place on the campus, including the lavatories. In fact cleaning of lavatories had the first priority. It is particularly important in India, because cleaning of lavatories has been considered the work of untouchables. When I experienced that aspect of *nayee talim* I was reminded of Pestalozzi, who said that he descended down to their level (of the poor uneducated) so that he could lift them up with him. A very moving and uplifting experience indeed it was.

Actually, all the staff of the school and other members of the Sevagram community had made cleanliness a science as well as an art, developing its technique thoroughly. It became a separate subject in the school and the community. After those forty-five minutes of community cleaning everybody went to his or her work — some to the farm in which we tried to produce all the grains, vegetables and fruits needed for the community, others to their spinning and weaving workshops, and yet others to other activities, which included lunch for the community. We all took turns preparing dinner everyday. The number of workshops gradually increased according to the needs of the community and the growing educational programme.

After a couple of hours' break at midday all the children went to their studies related to the day's work and to other aspects of their life, such as health, culture, etc. In the afternoon there were music and art classes for all the children. Some of the time in the afternoon was also devoted to practice and rehearsals for the celebration of festivals and special occasions. Children and class teachers along with the art teacher decorated the places of celebrations and festivals with *alpana*, flowers and textiles. Stage craft and management was an important part of *nayee talim*. We had arrangements for simple sports and games — mostly games that needed no equipment.

Yet another feature that ought to be mentioned here is the character of festivals celebrated and the daily prayer conducted in *nayee talim* schools. Festivals of all religions were celebrated in as good a way as possible. School prayers were the same as those of Gandhiji's *ashram*, i.e., inclusive of all religions. It had a very positive impact on the children about the equality of all religions.

At the end it may be useful to mention here something about self-evaluation by students. A list of headings of paragraphs from the annual self-evaluation by a post-basic student would give an idea of the content of the education he received. Instead of the usual type of examinations the Sevagram school considered self-evaluation — along with evaluation by teachers — a more effective and creative method of judging the all-round progress of a student.

Some of the sub-titles from it were: My health; Social Life; Kitchen work; Community cleanliness; Prayers; Looking after guests; Looking after the sick; Festivals; Cooperative store; My behaviour in the hostel and in my family; Village work; Craft and self-sufficiency; Spinning and self-sufficiency in clothing; Building work; Gardening; Work in the printing press; Language and literature; Mother tongue; National language; Other languages (English); Industrial and agricultural science; Social science; Self study; My aptitude and future studies/work.

There is much to tell and explain about the experiment of *nayee talim*. It is impossible to put even a tenth of it in such a small space. There is much, though not sufficient, material that is available for further study about the subject. What is lacking is an analytical study of *nayee talim*. For example, a question is often asked: In spite of *nayee talim* being such an ideal and practical system of education, why did it not spread as it should have? One day a thorough study on this question will have to be done. For the present purpose, I hope the above should suffice.

13

Development of Skills and Attitudes in Basic Education

Parthasarathi Banerjee

POPULAR basic education is becoming more and more elusive. This unattainability is reflected in its failure to encompass the mass of people, in its inability to offer a meaningful pedagogy and in its inability to offer to the people a satisfying instrument. It is amazing that not only a populous country such as India, or a less populous but perhaps poorer country as one would find in Africa, have failed in providing for education; but also the richer countries that boast of statistics of 'literate' populace sadly lack an educated mass of people. As a result the edifice of 'public space', built up through literacy or informative institutions of popular education and on which the pillars of democracy were mounted, is now crumbling. Ironically, neither 'science' nor 'democracy' appears to be in a position to confront this fear of death.

Naturally, for some proponents of science, it is time to redefine and rechart their enterprise. It is equally so for the proponents of democracy or of the 'public sphere'. Such a strategy may pay some dividends to those who want to save their own houses. However, it is unlikely to better the lives of the millions.

Surely our central inquiry cannot set aside the question of conducting a good life — a life that can satisfy its desires through undertaking enlightened actions. If this is our concern, how the millions can be enabled to live a good life, then surely the other concerns regarding democracy, science, theory or literacy will have to abide by this centrality or will have to be given up. Is it not legitimate to pay central attention to this question of basic education, how the people can be enabled to conduct a life which assures satisfaction?

Let us begin by asking the simple question, what is primary education for? Is it in order to inculcate the 'spirit' or 'ideology' of a civilisation (compare with the *bildung* concept), or to empower with liberationist ideals, or to make each child 'literate'? May we, in contrast, conceive of primary education as the inculcation of an attitude (of truth) and the drilling of the student through an engagement with a productive skill? Our exploration is of this latter kind.

In basic education we argue for two basic elements. The first is drilling in a productive engagement, following authoritative rules, such that the student may acquire a skill. Second is the inculcation of an attitude of aesthetics or of truth so that the student may engage with the world later in life, fearlessly, in order to know and enjoy and also in order to secure the good for which she/he has acted. We also argue that the characteristic feature of basic education is that it is non-theoretical. The goal of non-theoretical basic education is to enable the student to undertake independent action.

This summary statement regarding basic education is derived from two sources. The first is the tradition of imparting basic education in India. The second is some distinctive features of the Indian theoretical tradition that provide the ground and the content of basic education. This tradition of conducting basic education is rather old. We shall, however, not refer to the history of this business. We shall instead refer mostly to some necessary elements of the theoretical tradition.

Divides in Education

What is the legitimate and appropriate basis of divisions in education? A 'primary' education is normally defined as the first tier of a segmented educational system. A segmented educational system is not only the result of historical evolution in Western Europe, but also cannot be defined either on grounds of epistemology or on grounds of practical reasoning or pragmatic concerns. As a result, 'primary' education, as existing in this segmented system of education, can neither lead the student to higher and more abstract levels of epistemic concerns nor can it lead the student to practical concerns of achieving the good in life.

There is now a likelihood of dividing the educational system in terms of its exposure to information and its ability to process information. The primary aim of this approach, it appears, is to replace skill and knowledge with an allegedly 'objective' status describable in terms of information. With the advent of new technologies of handling information, and also with the growing disintegration of society, this informational basis of education is gaining ground. As of now, the educational system cannot be divided on this ground; however, there remains a distinct trend in that direction.

Contemporary slogans around literacy are largely reducible to such a classification.

Exposure to information and logical operations on that constitute the basis of divisions. Instead of going into debates, it may be briefly pointed out that a load of information is not knowledge; neither is a skill information processing capacity. Moreover, non-numeric information has to be described or literally narrated. Finally, a wish or a desire cannot be reduced to information. As a result, primary education cannot be defined on the basis of its exposure to information. Sadly, however, the latest pedagogy is being reduced to the provision of structured information. Organised money, a politics of hegemony and a group of opportunist ideologues are doing their best to enforce on the child this information-derived pedagogy, which cannot be justified on grounds of rationality and aesthetics.

This first tier of education can also not be defined by the primacy of a set of certain text-books. Proponents of such a view define segmentation in terms of interpretive capability. In fact, the literacy view of education also proposes this iconic relevance of textual signs. 'Texts' and 'text-books' assume a major significance. 'Connotations of civilisation' (*bildung*), 'conscience' and morality too are attached to this view. However, by being limited to interpretation, education here limits its capacity for inferential reasoning, its materiality and its activities on transforming the material. Moreover, such a view attaches undue importance to teleology and culture. Hence this basis of education and educational segmentation is also not acceptable.

Significantly, Indian educational practice is very different from these groups. However, in contemporary India, especially in urban areas, we have a significant presence of the institutions of these three above groups. Governmental activities are largely reducible to these practices. In contrast, the history of educational practice in India and the debates and practices on education since the middle of the nineteenth century to the middle of this one, especially the practice and thinking of Gandhi, Tagore, Bankim, Vinoba and others do provide a strikingly different alternative. In fact, in this essay, what we argue about is a summary only of this alternative Indian practice.

Facets of Indian Practice

Knowledge, according to Indian theorists, must have its characteristic concept and a characteristic good that is realisable. This concept must specify not only what it is but also what it is not. This unambiguity in the concept would not allow it to get mixed up with another concept. Separations are inviolable and are due to distinct epistemological departures. Corresponding to four types of epistemology there are four distinct knowledge-departures or systems (*prasthan*). We are concerned here with two knowledge-departures, viz., the knowledge-system pertaining to politics (*arthashastra*), which is distinct from the knowledge-system pertaining to wealth, economics and society (*vartashastra*). The other characteristic of any knowledge-system is the good that one may realise upon practising its concepts.

Each knowledge-departure must have its characteristic good. There are, however, pluralities of good and practices, most of which cannot be immediately recognised as belonging to the good of some particular knowledge-system. These practices therefore give rise to arts and skills (*vidya*, a subdivision or a particular practice). The concepts of any knowledge-system, according to the theorists, must engage in dialogues with the *vidya* practices, in order first, that the concept acquires its validating example from the common practice of *vidya*, and second, that the *vidya* in turn receives its conceptual or theoretical backing. This dialogue therefore is the first basis of the institution of acquiring knowledge and skill. The second basis of this institution is provided by the dialogues between the contending concepts of a knowledge-departure. A group of such concepts, logically coherent and cosmologically meaningful, is called a *shastra* (a pedagogy).

The second type of dialogue is conducted between competing pedagogues. The third basis of the knowledge-institution is provided by a third type of dialogue, conducted between the teacher and the student. The teacher answers the queries of the student. These dialogues are held in a medium of attitude. This attitude relates to aesthetics — the state of truth. Life, queries and practices of knowledge-seeking, defending or refuting must correspond to this attitude of truth.

Knowledge is required in order to resolve doubts and thus in order to act meaningfully. Indian theorists argue for a common knowledge that is distinguishable from common sense beliefs. Such knowledge is obtained through iterated fruitful actions, through the authority of sentences (or words). Therefore the actions in a commonly led daily life are both meaningful and knowledge-driven. We argue for four sources of validation, viz., sentence, inference, direct perception and analogy. Moreover, as argued above, knowledge is entailed not by way of justification as such, but by the realisation of good or fruit-ladenness of meaning and actions or iterated actions.

We summarise the arguments put forth by Indian theoreticians on the subject of the description and divisions of knowledge as follows. Knowledge cannot be described by descriptions (or descriptive sentences) alone. Hence an attempt to describe or limit knowledge by theories — where theories are nothing but descriptive sentences, is incomplete. The folk or common experience of doing things successfully is an equal partner of knowledge. Knowledge is also described by naming and achieved through iterated practice. Moreover, such a complete picture of knowledge is always derivable through an injunction. Such an injunction may not always be clearly visible or discernible, but it can surely be inferred. Indian theoreticians thus argue for a different scheme of knowledge division. The argued positions always refer to a source of injunction, an injunction, a theory as a description, a set of names, and an iterated practice or action orientation. Theoretical or descriptive knowledge must therefore relate itself to the examples of realised good in common life through dialogue with common people.

These theorists argue that even an art or practice or skill may have a theory or may be theoretical. This set of four sentences or four types of knowledge describe knowledge in its entirety. An art, while being performing or iterative, would also require injunctions in order to delimit the performance and to remain correct. Similarly a theory would require, apart from the descriptive propositions, an injunction to limit the iterated performance (for example, the experiment seeking evidences) on the nominal (the characteristic name, for example, of the biological organism, on which the experimental performance is being carried out), through an iterated or algorithmic procedure, etc. However, while a theory pays strict attention to the descriptive propositions, an art or skill may pay stricter attention to the procedure or to the injunction or even to the characteristics that are nominal.

A *vidya* is defined differently. It is defined by virtue of the good that is entailed to the performer. *Vidya* is classified according to the destination. Some treatises define sixty-four, others define sixteen types of *vidya,* according to the particular good that it allows the performer to attain by way of exercising or performing through the concepts as laid out in that particular pedagogy. This classification tacitly assumes multiple forms of good, and defines a pedagogy according to the variabilities of good sought. Necessarily, concepts in this *vidya* are laid out in any of the four types of sentences described above. A pedagogy is defined in terms of a good in a such way that any performance on the pedagogic skill would necessarily involve the performer, the object of performance, the good to be attained, and above all a social milieu where the pedagogy, the concepts and the good are to be either defended or the opponent's views and performances are to be refuted, through entering into a discourse, dialogue or debate.

Basic Education is not Primary

In this tradition of educational practice, education is divided according to whether it is theoretical or non-theoretical. Since such a division does not refer to a hierarchy, Gandhi and Tagore preferred the use of the term **basic education**. A non-theoretical education is basic. The goal of a basic education is to enable a student to acquire the desired fruit through his or her own actions. Basic education not only enables a person, by imparting a skill, to undertake the right action, but also provides for the meaning or the acquisition of the fruit of that action. Moreover, in case the skill fails the person's reasoning, basic education provides for the inculcation of an attitude of truth.

Non-basic or higher education is primarily theoretical. Its emphases are on inferencing, theory-defending and theory-refuting. This emphasis on theory is due to the presence of a situation of discourse. The goal of higher education is conceptualisation, defending one's own concept and refuting the opponent's concept. The instrument of these activities is theory. Basic education, in contrast, is not engaged in building, defending or refuting theories. It is therefore not in a situation of discourse.

The defining characteristic of basic education is skill formation and attitude development in a non-discourse situation. This education is aimed at the conducting of a good practical life. It is thus not primary. It can be provided to any age group or to anyone having the desire to conduct a practical and pragmatic life.

Basic education too is describable in four sentences. In basic education too we require a source of injunction (the *guru*), an injunction (on what not to undertake), a causal description in the form of a theory, a set of names of the terms for which the action has been undertaken, and a set of rule-based integrated practices or action orientation. Indian practice in basic education puts great emphasis on some of these four types. Basic education has the goal of imparting a *vidya*. A *vidya* offers a realisable good. It is attained and maintained in a non-discourse environment.

Skill in Object Transformation

The conducting of a good practical life requires skill in material transformation. Acquiring such a skill is the primary goal of basic education. It must be recognised that this formulation of our preceptors, including Gandhi and Tagore, is not only practical and meant for the good of the broadest mass of people, but is also rational. Justification of this formulation is possible on grounds of politics as well as rationality.

We may consider the *charkha* and pottery as examples of productive engagement. Such an engagement provides the student with (a) a concrete material at hand, which is the basis for later empirical investigation or action, (b) meanings associated with the following of authoritative injunctions and optative statements or even rules, (c) iterated practice that provides the skill — a cognitive and bodily endowment not reducible to information, and (d) knowledge of the particular.

In fact the objective of engaging a child in a productive engagement is not the development of a productive skill. The development of a productive skill is undertaken at an age when the child may enter the profession as a probationer. The child's encounter with the material nominal through an engagement with the productive set-up, such as in the preparation of clay or in the mastering of turning the potter's wheel, lays the foundation for a bodily and cognitive endowment. Such an endowment alone can provide the child-student with (a) those spheres of feeling that provide for the cognitive reckoning of shapes, directions, number, etc.; (b) those spheres of transformation that provide for the bodily reckoning of the transformability of material, iterated actions, etc.; (c) those spheres of reasoning that provide for inference, causal descriptions, etc.; (d) those spheres of search that look for validation, if not in inference then at least in the cosmology or in the conducts laid out in the *itihasa*; and (e) the relationship between desire, meaning and action, such that the child can associate action with the realisation of a good, the satisfaction of desire, providing not only a valid meaning but also the grounds of rationality in action.

In short, the child-student's productive engagement with the material nominal provides the rational basis for action. A skilled action is the next state. Basic education provides the foundation for the development of skill. Basic education in the first few years may or may not impart a specific skill. It should, however, give the student the necessary rigorous drilling — continuous and iterated action, often following rules or authoritative prescriptions. A specific productive skill may be imparted to the grown-up student either at the premises of the school or at the premises of a production set-up. The first few years of basic education provide the foundation for this skill-building. The goal of basic education is to attain a *vidya* describable in terms of a realisable good.

Skill Institutions and Public Space

Vartashastra, we may recall from our earlier discussion, is about those concepts that deal with wealth generation, trade and commerce, manufacturing, social exchanges — in short, much of the contemporary disciplines that are devoted to the sciences of nature and life, engineering and technology, economics and sociology. This pedagogy, as we have argued above, is after those goods which by being related to wealth, commerce, economy, etc., offer rules, algorithms, etc. for the practice of the arts and skills in such a way as to secure a validation following the performance of the *vidya*. Therefore, the realisable good being known, the *vidya* would ensure a valid concept only following the undertaking of the performance. Hence the institutional features should ensure the inculcation of a theoretical or algorithmic underpinning. Second, the *vidyas*, not being in need of prior validation, would not require a generalised epistemic or 'scientific', i.e., abstract, conceptual background. Third, these *vidyas* should be primarily either descriptive or algorithmic, hence value or normativeness should not be of much concern. Fourth, these *vidyas* would not require, in general. dialogues in refutation, but would mostly depend on authority and dialogue for knowledge — hence apprenticeship. Fifth, as a result and since the good of the practical life is known, these *vidyas* would not require 'specialists' and 'theoreticians' to sermonise on the normative. Sixth, these *vidyas* are designed as though to cope with the vicissitudes of uncertainty, plausibility and multiplicity of values.

It has been conjectured, for example by Bankim, that social and pedagogic institutions in India of the pre-British period were harmonious, and therefore even in its lived worlds the theme of that harmony could be accepted institutionally, at least insofar as the authority of the sovereign was willing to achieve that. Therefore *pathshalas* were not just 'elementary' schools but primarily instructional institutions in the practice of the *vidyas* of the *vartashastra*. Such instructional schools were in close cooperation with the apprenticeship of the students in parts of the lived-in world, such as forging shops or agriculture or trade. Colonisation brought an end to this state of affairs, and by about the 1830s *pathshalas* were reduced to only elementary schools. By the middle of the nineteenth century the rural population had lost the knowledge of rules and algorithms. What remained at the turn of the century were a greatly impoverished rural people with no public space left for the attaining of good in practical life or for conducting a knowledgeable dialogue. Daily life in villages had become vicious. Many sensitive minds wrote about the deplorable state of rural Bengal.

Public spaces impregnated with knowledge used to be the norm. The practical life led by the *vidyas* of the *vartashastra*, as it was limited to the descriptive, could find cosmological and thematised expression through the rather regular narratives of the narrators (*kathak*), delivered as aesthetically as possible. By about the 1850s, slowly the narrators too were to leave the scene. Bankimchandra lamented this in his essay on public education (Bankim 1952: 377), and surely could see that newspapers were not an alternative. The other mode of public space, that had become largely restricted by the eighteenth century, and was to vanish by the third quarter of the nineteenth century, was the institution of public dialogue in refutation (*vichar sabha*) on the concepts and resolutions of the pedagogues. Quite importantly, it was not only the esoteric and the abstract reasoning of the contending pedagogues that used to draw the attention of learned minds; often such dialogues were conducted on important social and ethical problems. In fact the extent to which Rammohun Roy, Dayananda Sarasvati or Vidyasagar could draw the attention of both English learned minds and traditional scholars in debates on the relevance of monism or of the *Veda* or regarding widow remarriage, etc., was possible only because of the remaining impressions of the earlier institutions of a knowledgeable public sphere. With the demise of the *tol*, this institution too first lost its importance and then ceased to exist.

Some Events from Recent History

Rabindranath was one to understand early the limits of the 'nationalistic' approach to education. Aurobindo defined the nation as *dharma* (disposition of acting under injunctions and knowledge); Tilak defined it through the space of action; Swami Vivekananda and Swami Dayananda defined the national awakening through disenchanted actions and through action-performance (*yajna*) for the achievement of a real; Bankimchandra defined it through devotional practice and Bepinchandra too through such devotional practice (Rabindranath 1961; Bankimchandra 1954; Bepin 1954a; Bepin 1954b, 1957; Vivekananda 1981). Rabindranath rejected the Western notion of the nation and declared categorically the meaninglessness of a 'national' approach to the reconstruction of India. To each of them, importance was to be attached to the power of the self acquired through practice. Rabindranath, in a series of essays, dwelt on the power of the self and that of the collective, and rejected any emphasis on the latter. The collective, he argued, was to be attained through the attainment of the good of the self. The good of the self is in knowledge, as much of it is in the seeking of the universal so much of it is in the reality of life — in its particularities and in its movement towards a better and enlightened life.

This was the period when Rabindranath went to the villages in order to examine the extent and causes of poverty, paucity of drinking water, ignorance and ill health. He came closest to the traditional understanding and realised the necessity of the institutions, depicted above in the last section, that would bridge, through knowledge and practice, the existing gap between pure and common knowledge. The former he conceptualised as the Vishva-Bharati (the abode of universal knowledge) and the latter as the Sri-Niketan (the beautiful dwelling). In this experiment, life in the dwelling is in tune with the quest for the knowledge beyond, and the knowledge of the beyond gets its form in the practical performances of the dwelling. Sri-Niketan was thus for the common villagers. It was to bring advanced agricultural engineering and practices, irrigation and potable water, craft skills and healthy practical life, and above all education (even through evening classes for adults) to them. Vishva-Bharati was virtually in the same village. It was the centre of higher learning. It was not to follow, logically, curricula and contoured content as laid down by a discipline. The dialogue in knowledge, as its motto, was to be conducted in the open, not barred from the public. The other dialogue, with the common and the practical life, was to be necessarily in the open — in the public space, through such congregations as a *mela* (fair).

Rabindranath was not alone. Many, young and old, gathered around him. More important was the spread of the idea. Schools or centres came up, or existing institutions were transformed, in Assam, Bihar, Punjab, Gujarat, Maharashtra, Tamil Nadu. The sources of the idea were many and the nature of experiments too was varied. Kaka Kalelkar and Satish Dasgupta were surely to be considered exemplary.

For many practitioners the other profound fountainhead was Mahatma Gandhi. Gandhi's insistence was on the conducting of a healthy practical life as the first and abiding minimum of any education.

> Charkha and the removal of untouchability are two of the most essential ingredients of our education. (Gandhi 1962:96)

> Every teacher must learn, if he has not already learnt, spinning, carding, ginning and determining the kind and quality of yarn . . . . Just as we all eat and know how to put on clothes, even so we must know spinning, etc. . . . . We want to put some flesh on the human skeletons to be found [in] our villages throughout India. (*ibid.*, p. 97)

Further,

> Charkha is the symbol of labour. When it is firmly established in its place, carpentry and smithy, etc., will come in automatically. (*ibid.*, p.114)

Skilled living is the language of life. Gandhi therefore brought to skilled living and the employment of the mother tongue an eminence in the entire scheme of learning.

> We have paid dearly for having all these years learnt everything through the medium of English. We have strayed from the path of duty. Then take the teaching of economics. The present system obtaining in government institutions is vicious. Each country has its own economics. German textbooks are different from the English. Free trade may be England's salvation. It spells our ruin. We have yet to formulate a system of Indian economics. The same about history. Here . . ., therefore, there is a vast field for you and your teachers for original research. . . . Then we are putting a special emphasis on manual and industrial training. . . . It is not by making our brains a storehouse for cramming facts that our understanding is opened. An intelligent approach to an industrial training is often a more valuable aid to the intellect than an indifferent reading of literature (*ibid.*, pp.145-46).

It is thus important to understand that each *vidya*, for Gandhi and for a very large number of other Indians, was to be defined by the characteristic good and also by characteristic concepts. Gandhi had said quite categorically,

> I am not opposed to education even of the highest type attainable in the world. The state must pay for it wherever it has definite use for it. I am opposed to all Higher Education being paid for from the general revenue. It is my firm conviction that the vast amount of the so-called education in

> arts given in our colleges is sheer waste. The medium of a foreign language through which Higher Education has been imparted in India has caused incalculable intellectual and moral injury to the nation. (*ibid.*, p.55)

He and others such as Vinoba Bhave argued that education had to be self-sustaining. The crucial point meant by this is that education has to be for a realisable good alone. He said,

> There is no insurmountable barrier between what is called intellectual education and the crafts . . . . Training in arts and crafts offers full scope for the development of the intellect. And I venture to claim that without it the development of intellect is impossible. (*ibid.*, p.159)

The famed Wardha Educational Conference of 1937 had the following among other resolutions:

> Higher Education should be left to private enterprise and for meeting national requirements whether in the various industries, technical arts, belles-lettres or fine arts. The State Universities should be purely examining bodies, self-supporting . . . . Universities will look after the whole of the field of education and will prepare and approve courses of studies in the various departments of education . . . . University charters should be given liberally to any body of persons of proved worth and integrity . . . . (*ibid.*, p. 297)

The signatories to the Wardha Declaration included Zakir Hussain, Aryanayakam, Vinoba Bhave, Kakasaheb Kalelkar and K.T. Shah.

This was a remarkable scheme. It was very close to what the Indian epistemic or validational approach to divisions of pedagogies, development of *vidyas* and their institutionalisation would have called for in a contemporary setting. The modernity of this approach was in its deliberate attempt at creating a public space for dialogue in knowledge. Quite remarkably, such a space was not to be institutionalised around the university. It was meant to be popular and to be outside arbitration by the state. Equally important it was that the pedagogies were considered in such a way as to obtain their evidential base from common, contemporary practical life and also, *inter alia*, to sustain the transformation of the art of the practical life into a *vidya* by way of enriched feedback from an evidentially enriched pedagogy. This was therefore the second dimension of public space sustained through the interactions between the pedagogies and the *vidyas*. Pedagogies, for their finer enrichment, were to be primarily supported privately, and only in cases of national requirement would the state's endorsement be considered necessary. Hence a 'disciplinisation' through the enforcement of disciplines was entirely rejected. Finally, as his comments about economics show, he was convinced that each *vidya* and pedagogy was meant to secure a good of its own. This good, if it was to be part of the *vartashastra*, was surely to respond to the contemporary state of affairs in the world. Perhaps the only aspect that he did not specifically speak upon was the validational and epistemic departures of pedagogies.

There were others, however, especially from among scholars in Sanskrit, who categorically emphasised this point. In Bengal alone three great scholars responded. Mm. Phanibhusan Tarkavagis brought out in Bengali a detailed commentary along with the commentaries of Vatsyayana or the *Nyaya* aphorisms. Mm. Yogendranath Bagchi brought out several texts, often along with commentaries on the validational-epistemological basis of knowledge, as well as the necessary conditionality on the knowledge to have its own good. Pt. Pancanan Tarkaratna brought out annotated *Puranas* and other commentaries on such texts as would relate to good-pleasure, norms and ought, etc.

The emphasis they brought was both far-sighted and of contemporary concern, because they wanted the debate to take place around the pedagogical division of knowledge and its relation of sustenance to various *vidyas*, the relation of *vidyas* to practical life and good, the crucial importance of dialogues of various types, etc. Similar efforts were noticeable also at other places in India, notably at Banaras. Some other scholars and political leaders too joined this debate. Amongst the first would be such philosophers as Krishnachandra Bhattacharya (he contributed a remarkable prose on 'Swaraj in Ideas'); or Radhakrishnan, or the sociologist Radhakamal Mukherjee, or the sociologist-economist Benoy Kumar Sarkar. Amongst the political leaders Bepin Pal (for example, through reinterpreting Bengal Vaishnavism in 1933) or Aurobindo Ghosh (through several writings) were surely not alone.

This was a debate of importance. What is noteworthy is that the debate was reconstructed variously by the state machineries, the media and political parties in such a way as to disregard the theoretical dialogues of the Sanskrit scholars. The obvious result was that Gandhi (and also Rabindranath, although he had differences with Gandhi) could be reduced to allegedly non-modern positions, and the debate could be shifted to a wrongly posed question about modernity (understood as contemporaneity, or positive to the West) *vis-a-vis* its antinomy. The constructors of this false reading of the debate finally reaped the advantage by pushing away not only the aspirations of Sanskrit scholars but also those of Gandhi and Rabindranath. The advantages were reflected in the further spread of universities and colleges, strict enforcement of 'disciplines' and discipline-based professionalisation, through urbanisation and deepening of the dependency on English in especially higher education, decimation of the knowledgeable public and pedagogical dialogues, especially from small towns and villages, conversion of dialogues into media popularisation of 'issues', near-total dependency on Anglo-American universities for higher learning, etc. They were reduced to this virtual antinomian position to Gandhi who had spoken, in contrast, about a broader base of higher education and research in such areas as sciences, engineering, economics, other social and cultural studies.

The names of Benoy Sarkar and Srinivasa Sastri are worth mentioning in this regard. Sarkar had a creative engagement with education: since 1907 he was one of the architects of national education. Early in his life he established about a dozen national schools in the districts of Malda and Dacca, joined the N.C.E. of Calcutta as a professor, wrote on the science of education and on economics, sociology, history. He was perhaps one of the most able organisers and a gifted writer. In 1940 he published one of his public lectures in the *Calcutta Review* (June 1940) under the title 'Sociology of creative disequilibrium in education'. He said:

> Discontent, disharmony, disequilibrium and so forth in matters educational as in other spheres have to be accepted as the eternal and universal items in the individual psyche as well as in inter-human or societal relations. Every so-called synthesis is in reality a condition of conflict or disequilibrium. And virtually in every instance this disequilibrium is creative and evocative of fresh values. Not to be prepared for such evolutive discontent or creative disequilibrium should be treated as the worst disqualification for educational statesmanship or cultural patriotism . . . .

Sarkar's thesis was thus in clear opposition to the views of both Gandhi and even partially of Rabindranath. Sarkar also had argued, in the same essay, not to look to state aid in the first instance or in the second instance, but to energize independently and strive individually as often and as long as possible without support from the governmental authorities. Presumably, to Sarkar Gandhi's Wardha scheme was pedagogically unsound as based on an incomplete and inadequate recognition of the diverse demands of the individual personality and the full-blooded 'creativity' of human beings (Chaudhury 1940: 24).

Sarkar's opposition to Gandhi was thus based on creative disequilibriation in each individual; *swaraj* or autonomy of the literary, scientific, and aesthetic; no *vidya* or *kala* is so singularly fundamental that other *vidyas* can be sustained by that alone. It is important to recognise that these differences are based on arguments that are deep indeed; many contemporary European critics are also trying to reformulate the scheme of knowledge-division similarly. Sarkar at the same time knew and agreed with Gandhi's pragmatism. He argued that for educating the overwhelming majority of the Indian population the Wardha scheme could be trusted.

Social Consent and Object Transformation

The above provides the rational basis of action to the student as an individual. Does the same education also generate a sense of social obligation? In fact, this basic education on productive and skilling engagement provides twin bases of social cooperation.

The first base is owing to the teacher-student relationship. This is a relation of initiator-initiated or instructor-instructed. The statements of an initiator are action sentences. Such sentences should not be vitiated by any self-interest on the part of the instructor. Understandably a basic instructor is more likely to qualify for such an attitude compared with an instructor meant for on-the-job training to grown-up students. The instructional utterances of a basic instructor initiate the child student. The child student clearly recognises the importance of this authoritative source, which can issue injunctional or optative utterances preferably not vitiated by self-interest. This gives rise to the recognition of an authority.

The instruction of the instructor is often in the form of a rule, recalled from memory or from tradition. The child-student clearly recognises not only the continuity of a lineage but also the endowment bequeathed by heritage. It is a debt, therefore, that has to be necessarily met. Basic education and the action of the child necessarily should give recognition to this debt and one's own obligation to satisfy it.

It is perhaps even more important that the student is taught the importance of the consent of others to the action being undertaken by him. Any action, with its goal, must be declared by the undertaker, and consent to this undertaking be sought beforehand. Such consent, it may be argued, is better than that based on such principles as the principle of no-harm. This is also non-normative. Consent-seeking, being clearly non-moral, seeks to arrive at a judgement on the future. This is a superior principle of social action. However, it does not provide the bases for collaborative action. Action-orientation is one of the primary goals of a basic education. This orientation, according to Tagore, Gandhi or the other preceptors, is individualistic. The objective of basic education is to orient the individual student to undertake one's own action. In order to undertake and prior to commencing one's own action, the student must fulfil his or her debt obligation and seek consent.

The Question of Attitude

Basic education is not limited to productive engagement. It has two more elements. First is the fabulous historical or descriptive narratives (the *itihasa*), and the second is responsible cooperative social engagement. These together provide for the inculcation of the attitude of truth. Such an attitude alone can lead the student from the knowledge of a particular to that of numerous particulars.

Narratives are fascinating. While iterated practice on, say, the *charkha* gives shape to the mind or forms the mind, the narratives provide this formed mind with both a memory of the lineages and a future of the imagination. Mind-formation or skilling takes place through the engagement of the senses with the particular material (e.g., clay or cotton). Narration disengages the mind from senses and action. It takes the mind to the beyond-sense of lineages, memory, or the future.

These historical narratives and those on lineages or on cosmology, largely replace the role of theoretical discourse in non-basic education with fabulous narrations. Skilled practice and the transformation of objects are but little known theoretically. Basic education need not theorise on this transformational practice; neither does it need to be defended against an alternative practice. In fact value-pluralities and rule-pluralities are welcomed in basic education. The good of a *vidya* is unique, while action peculiarities are non-unique or plural. Basic education nurtures a respect and a dignity for the plural. However, what basic education requires is an attitude towards useful action in an unknown domain. Such an attitude, needing imagination and memory, is provided by the fabulous narratives.

Responsible social cooperation again is the admission that the young student has a duty. The Gandhi-Tagore ideas and practice on duty are derived not from moral penance nor from an authority, but from an attitude towards obligation and towards the consent of others. This obligation cannot be reasoned out, and hence the student is to look to tradition. Moreover, the student is supposed to get the consent of others prior to undertaking an action. The student develops an attitude towards cooperative work through seeking the consent of others.

One source of knowledge about one's obligations is the conduct of the preceptors. In its absence, cosmological or historical narrations on conduct appear significant. Therefore basic education does not put an emphasis on moral teaching or on the 'conscientisation' of the pupil. In contrast, it offers a pragmatic ground, through the lived life of the preceptors, to the attitude towards obligation.

Most important, perhaps, is the significance of *abhinaya* and *nataka* (close equivalents are perhaps playing and drama) in basic education. We cannot, however, elaborate on this theme here.

The aesthetic elements here are derived from a taste of the performances that one has undertaken. This aesthetic is not derived from the goal of the action; on the contrary it is derived from the taste of the performance. Interestingly the student learns that life, roles in life and conduct in life are not damningly serious. The consternation that one faces in a contemporary school, where the children are shown a life of deadening commitment, is indeed to be laughed at. Indian basic education is based on playfulness and an aesthetic of tasting the performances. Knowledge, according to Indian theory, is not based on the intersubjective world. The intersubjective world is either for a discourse or for *abhinaya*. Such a world has to be lived in aesthetically, i.e., with an attitude and through playing. Since basic education provides a non-discourse environment, it puts great emphasis on the aesthetic and the playfulness of *abhinaya* and *nataka*. Rabindranath designed basic education on this line. Bankim theoretically argued for it, and Gandhi thought that it was inseparable from basic education.

Cooperative work is based on cooperation between a few role-performances. A minimum of four roles are conceived. The first role is to issue authoritative instruction, based primarily on rules of performance. The second role is to describe or delineate the causal connectives, to infer, etc. The third role is to make use of narratives and to imagine. The fourth role is to put together these three and act.

The descriptive parts of the narration trains the student to understand causal relations. Such training gives the mind a capacity to investigate and enquire, to infer and to lay bare causal associations. Much of this is known as science in contemporary jargon.

Recalling the student's engagement with the *charkha* (or cotton), from whence he acquires the bodily-cognitive skill, we may now understand that the causal description and the fabulous history enable him to undertake independent action. This is the goal of non-theoretical basic education. The goal is to enable the student to acquire skill and to undertake action independently that is perhaps partly known to him descriptively (or scientifically) and partly known to him as narration from memory or tradition. This unknowable, or the unknown, can be acted upon by the student only if he/she has an attitude of truth and the consent of others. Such a consensual situation does not allow the student to initiate action that could be harmful to living organisms and to the tradition.

This provides us with a pedagogy and the goal of basic education, which form the summary of experience in conducting a variety of experimental basic schools by our preceptors. These schools have followed the ideas of Gandhi and Tagore and of tradition. Our attempt is to re-emphasise the importance of productive engagement and of attitude in non-theoretical basic education.

Debates on Attitude

Radhakrishnan said in 1961 to teachers:

> We in our country look upon teachers as gurus, acharyas. What do these words indicate? Acharya is one whose achar or conduct is exemplary, is good. He who is able to remove that kind of spiritual blindness is called guru . . . (Radhakrishnan 1963:166)

> The way to develop a civilized community is by developing understanding and compassion. Inward wisdom and outward compassion. If you become merely a saksara — a literate man — and you do not have the moral principles and you do not cultivate wisdom. you will become a raksasa . . . mad with power, intoxicated with might, trying not to build up a world but to wreck the world. (*ibid.*, pp.168-69)

He thought that the university was unity — unity between the mundane and the universal, people with learning, between contending concepts, between knowledge and the *vidyas*. Beginning with Rabindranath and Gandhi, this idea flowed past: the Kothari Commission of 1964 too spoke upon the social nature of the university, the vocation of art-skill (*vidya*), the difference between literacy and education.

The attitude of truth (*tarka*) is indispensable to an inference. Gandhi, Rabindranath, Aurobindo or Ambedkar, Jotiba Phule — to all, attitude and conduct constituted a necessity. A contemporary democracy cannot run on mere popular opinion; it requires to be suffused with a knowledgeable public dialogue. And given the fact that contemporary India had lost its own earlier institutions, it was Gandhi's or Bepin Pal's task to impregnate the public space with an attitude without which no knowledge could have been accepted. Attitude (*vritti*) is amongst the oldest conceptual tools in Indian thought. A particular knowledge attained is an attitude, and one traditional approach is to conceive of the ultimate knowledge attainable as consisting of sixty-four *vrittis*.

Bankimchandra, one of the most influential thinkers of India, analysed education as *vritti* only. In his argument, a good too is defined similarly. Bankimchandra thought of three classes of attitudes, first for knowledge, second for performance and third for the aesthetic. Humanity can flourish in the fulfilment of the three classes of attitudes. He argued that the objective of education (cultivation) is to empower by way of allowing attitudes to flourish as well as by acquiring knowledge. Moreover, cultivation of attitudes of knowledge needs to be accompanied by the cultivation of the other classes of attitudes. The set of attitudes, in turn, seeks fulfilment in the virtues of compassion, friendliness, joyousness and indifference to others' frailty. This is also *dharma*. Love for others and a sense of duty to the community and to the country are derived from this compassion — the ultimate confluence in which all the attitudes are synthesised. Bankimchandra therefore imagined Krishna as the ideal.

since it is in him alone that one can see all the attitudes, knowledge, performance, aesthetic, compassion. The ideal, according to him, is in the ultimate fulfilment of humanity. It can and ought to be pursued through iterated practice.

Attitude is also therefore an index of power, a power which is not aggrandising. This is the position adopted by almost all Indian thinkers in the period we are concerned with. Such power is a-causal, as a sovereign or an authority is a-causal. This is what the Indian theory of sovereignty of state claims. The attitude therefore rules. Indian thinkers were organising the spectrum of knowledge in terms of an institution where power of attitude would rule the pedagogy, the *vidyas*, the *tantras*. The *pathshala*, the university of their conception, was to be in dialogue with the common and was not to be bounded by four walls and by the enforcement of disciplines.

The Institution of Basic Education

We may note a few distinctive features in the history of Indian education. Primary education was widespread even as late as the early nineteenth century. Often nearly every village would have a school, to which children would come from nearly all the population groups proportionately. Second, the curricula were the result of long practice as well as of its engagement with productive activities. There was neither any state financial support nor any control over curricula. Third, the public space of popular life and institutions was often constituted around (a) dialogue with theoretical education through varieties of institutions and fabulous historical narrations; (b) dialogue with productive activities; and (c) an attitude that was non-normative but one of forbearance, aesthetic and truth.

These unique features did not provide for a segmentation of education. The British colonial attempt was to replace this with segmented educational institutions and segmented educational content. Such attempts were resisted. Contemporary Indian education bespeaks this confrontation.

In this institution there are therefore two basic elements. First is a drilling in a productive engagement such that the student may acquire skill. Second is the inculcation of an attitude of aesthetic, of truth, so that the student can engage with the world later in life, fearlessly, in order to know and enjoy the realised good.

It is important also to recognise the institutional aspects of this approach. The distinctive features of the institution are that (a) it does not segment education in a hierarchy but divides knowledge and discourse epistemologically and therefore non-hierarchically; (b) it is not based on a divide between theory and practice and does not reduce science to a set of information — it brings knowledge to the common people; (c) science no longer remains in isolation but is created anew skilfully while the discourse with abstract theories of science continues through narration, description and an attitude of truth.

As a result, skilled practice receives its nourishment from advances in science. The student trainee becomes responsible to others in society. In short, this approach leads to an enlightened society. What is relevant socially and politically is that education becomes widespread, nourished by the material availability in its locale; it is maintained and nurtured by the local people, trained by local skill masters. It must be mentioned that this is most environment friendly as well. Such a basic education does not ideologise young minds into a civilisational mould, nor does it moralise (through freeing the conscience) or simply inform and attempt only to make students literate. In contrast, this basic non-theoretical education skills the population, nurtures aesthetic and an attitude of truth and of cooperative work.

References

Bagchi, Yogendranatha, 1947, *The theories of state in early India* (in Bengali). Calcutta: Pracyabani.

———, 1974, *Theory of validation according to early nyaya and early mimamsa.* Calcutta: Sanskrit Pustak Bhandar.

Banerjee, Parthasarathi, 1995, "Induction of 'social-sciences' as the disciplinary triplet: themes from Indian response". New Delhi: NISTADS (mimeo).

Bankimchandra, c.1952/1954, *Bankim rachanavali,* vol.II. Ed. by J.C Bagal (in Bengali). Calcutta: Sahitya Samsad.

Chandra, Bipan, 1966, *The rise and growth of economic nationalism in India.* New Delhi: People's Publishing House.

Gandhi, M.K., 1962, *The problem of education.* Ahmedabad: Navajivan Publishing House.

Pal, Bepin C., 1954a, *Writings and speeches.* vol.I, parts I & II. Calcutta: Yugayatri Prakashak Ltd.

———, 1954b, *Swadeshi & Swaraj.* Calcutta: Yugayatri Prakashak Ltd.

———, 1957 *Character sketches.* Calcutta: Yugayatri Prakashak Ltd.

Radhakrishnan, S., 1963, *Occasional speeches and writings.* New Delhi:Publications Division.

Rai, Lajpat, 1928, *Unhappy India.* Calcutta: Bauna Publishing Company.

Sahajananda Saraswati, 1985, *My life story* (in Hindi). New Delhi: People's Publishing House.

Tagore, Rabindranath, 1961, *Collected works.* vols. XII & XIII (in Bengali). Calcutta: West Bengal Government.

14

Indian Strategies to Achieve Universalisation of Elementary Education

K. Gopalan

FOUR and a half decades ago, we in India had taken a pledge through our Constitution that within a period of ten years from 1950, free and compulsory elementary education would be provided to all children up to 14 years of age. Since 1950, determined efforts were made towards the achievement of this goal. Over the years, there have been very impressive increases in the number and spread of institutions as well as enrolment. Today, India has about 574,000 primary schools (classes I-V) and 156,000 upper primary schools (classes VI-VIII), the number of teachers in them being 1.705 million and 1.082 million respectively. The enrolment at the primary and upper primary stages is 109 million and 40 million respectively. The Indian elementary education system is thus one of the biggest such systems in the world, providing accessibility within 1 km to over 825,000 habitations covering 94 per cent of the country's population. During the past one decade the enrolment rate has grown close to 100 per cent at the primary stage.

However, universalisation of elementary education (UEE) in its totality is still an elusive goal and much ground is yet to be covered. Drop-out rates continue to be high (36.3 per cent in classes I-V and 53 per cent in classes VI-VIII), retention of children in schools is poor, achievement levels are low, and wastage is considerable. Despite increased participation of girls, disparity still exists, more particularly among scheduled castes (SCs) and scheduled tribes (STs). In the National Policy on Education (NPE) 1986, with revised modifications in 1992, we resolved to achieve the goal of UEE by the turn of the century, emphasising three aspects: universal access and enrolment, universal retention up to 14 years of age, and a substantial improvement in the quality of education. The resolve is spelt out unequivocally and emphatically in the programme of action (POA) 1992, which gave unqualified priority to UEE. One is therefore pinning great hopes on the new innovations and alternative strategies which are being applied to ensure that the shortcomings and inadequacies, which did not allow us to realise this goal so far, are overcome, and the new resolve will not have to be extended further.

The Education for All (EFA) Summit of the nine high-population countries held in New Delhi in December 1993, which was an offshoot of the World Conference on Education for All held in Jomtien, Thailand, in March 1990, culminated in a policy declaration and framework of strategies for its implementation. The policy declaration called for providing basic education facilities for every child and consolidating efforts towards basic education for children, youth and adults. In the context of an integrated approach of basic education for all people, literacy and adult education programmes are to be improved and extended, eliminating disparities of access and improving the quality and relevance of basic education. It can be said that the Indian NPE 1986 and its POA 1992, while resolving to ensure free and compulsory education of satisfactory quality to all children up to 14 years of age by the year 2000, adumbrated the policy statement made at the EFA Summit.

The purpose of this paper is to briefly describe some of the new innovations and strategies that are being applied in India today to achieve UEE by the turn of this century.

New Innovations and Alternative Strategies

Some of the major initiatives and strategies are:

1. Disaggregated target setting and decentralised microplanning, which will provide the framework of universal access and community participation.
2. Strengthening alternative channels of schooling such as the non-formal education (NFE) system for those who cannot avail of conventional full-time schooling.
3. Introduction of minimum levels of learning (MLLs) at primary and upper primary stages to improve learner's achievement.
4. Improvement of school facilities by revamping the scheme of Operation Blackboard (OB) and connecting it to the MLL strategy.
5. Establishing linkages between programmes of early childhood care and education (ECCE), primary education, literacy and UEE.
6. Addressing the more difficult aspects of access, particularly to girls, disadvantaged groups and out-of-school children.

7. Restructuring of teacher training in view of the changed strategies and programmes.

8. Availing of external financial support for basic education.

9. Launching the National Elementary Education Mission (NEEM).

Disaggregated Target Getting and Decentralised Microplanning

Our experience with UEE encompasses the entire Third World experience. On the one hand, we have States like Kerala, which have achieved universal literacy as well as UEE in terms of school participation with social indicators as good as those of the best among the Third World countries. On the other, we have States like Bihar and Madhya Pradesh, with indicators as bad as those of the Sub-Saharan countries.

One of the new strategies to achieve UEE is adoption of disaggregated target setting and decentralised planning. Our long experience with the pursuit of UEE has established that UEE is contextual. The contextuality varies widely across the country. Even in a State like Kerala, where participation is near universal, much requires to be done in respect of quality and achievement. In such states, the pursuit of UEE would be mainly in the areas of quality, facilities and achievement, while in other states participation and demand aspects would need more attention. Therefore, the attempt would be to prepare district-specific and population-specific plans for UEE within the broad strategy frame of microplanning through people's participation. Microplanning has been defined as a family-wise and child-wise design of action to ensure that every child regularly attends a school or an NFE centre and completes 8 years of schooling at a pace suitable to him or her and attains an essential level of learning.

Guidelines for operationalising microplanning have been prepared and distributed to the state governments. The concepts of microplanning and local level capacity building have been given currency and efforts launched to decentralise educational planning and management. Microplanning exercises have already been undertaken in several states to ensure that all children receive primary education of satisfactory quality either through formal schools or through part-time NFE centres.

To operationalise the strategy for UEE through disaggregated target setting and decentralised planning, a new scheme titled District Primary Education Programme (DPEP) has been evolved. The concept of DPEP is a beachhead for effecting improvements and full-scale development of the entire elementary education sector. The overall goal of the programme is the reconstruction of primary education as a whole in the districts instead of piecemeal implementation of the various schemes. The fundamental principle of DPEP is capacity building at all levels to evolve further strategies which are replicable and sustainable. The specific objectives of the programme are:

- to reduce differences in enrolment, drop-out and learning achievement among gender and social groups to less than 5 per cent
- to reduce the overall primary drop-out rate to less than 10 per cent
- to raise average achievement levels by at least 25 per cent over measured baseline levels and ensure achievements of basic literacy and numeracy competencies and a minimum of 40 per cent achievement levels in other competencies by all primary school children,
- to provide, according to national norms, access for all children to primary education classes (I-V), i.e., primary schooling wherever possible or its equivalent non-formal education.

The programme would strengthen the capacity of national, state and district institutions and organisations for the planning, management and evaluation of primary education. NEEM has recently been set up to oversee, among other things, the implementation of this programme throughout the country.

Alternative Channels of Education such as the Non-Formal Education (NFE) System

Non-formal education has become an accepted alternative channel of education for children who cannot attend full-time schools due to various socio-economic constraints. To reach this large segment of marginalised children, we in India have been running, since 1979-80, a programme of NFE for children in the 6-14 age group, who have remained outside the formal system. These include drop-outs from formal schools, children from habitations without schools, working children, children who have to remain at home to do domestic chores, and girls who are unable to attend formal schools for a variety of reasons.

The enlarged and modified version of the NFE programme now in operation visualises NFE as a child-centred, environment-oriented and flexible system to meet the diverse educational needs of the geographically and socio-economically deprived sections of society. Non-formal education is designed to overcome the shortcomings of the formal school and make education a joyful activity. Decentralised community participation through village education committees (VECs) in planning, running and overseeing the programme has been considered crucial for its success. Although the focus of the programme is on the educationally backward states, it also covers urban slums and hilly tribal and desert areas in other states as well. Today, the programme is being implemented in 20 states and union territories through the state governments and voluntary organisations. While there are more than 226,000 NFE centres in the state sector, there are about 29,000 run by voluntary agencies. About 44 per cent of all the NFE centres are exclusively for girls, who are the main victims of socio-cultural and socio-economic factors. The estimated enrolment capacity is about 6.3 million children. Under the NFE programme, efforts are now being made to further improve quality, allow greater flexibility to implementing agencies and relocate NFE centres on the basis of microplanning/area survey. The NFE programme is being linked to ground realities, allowing for continuous experimentation. Development and scaling-up of effective NFE models that can help the learners to learn at their own pace is a major thrust area.

Minimum Levels of Learning (MLL)

The need to lay down minimum levels of learning (MLL) emerged from the basic concern that irrespective of caste, creed, location or sex, all children must be given access to education of a comparable standard. The MLL strategy is an attempt to combine quality with equity. It lays down learning outcomes in the form of competencies or levels of learning for each stage of elementary education. The strategy also prescribes adoption of measures that will ensure achievement of these levels by children both in formal schools and in NFE centres.

The focus of the MLL strategy is development of competency-based teaching and learning. Preliminary assessment of the existing levels of learning achievements has revealed that they are quite low across several districts. Minimum levels of learning in respect of three subjects, namely language, mathematics and environmental studies, have already been laid down for the primary stage. It has been stressed that the emphasis should be on concept formation rather than on content. The burdens of non-comprehension and overload of content are forcing children to resort to rote memorisation. The issues of content versus concept, understanding versus rote memorisation, unachievable content load versus achievable set of competencies, have been integrated into the new MLL approach. Minimum levels of learning have been specified in terms of competencies expected to be mastered by every child by the end of a particular class. The programme has been initiated throughout the country with the help of voluntary agencies, research institutions and others concerned. Minimum levels of learning for the upper primary stage are now being finalised.

Revamping the Scheme of Operation Blackboard (OB)

Recognising the unattractive school environment, unsatisfactory condition of school buildings, inadequate physical facilities, and insufficiency of instructional materials in primary schools, which function as demotivating factors for enrolment and retention, a scheme symbolically called Operation Blackboard was introduced in 1987-88 to bring all existing primary schools in the country to a minimum standard of physical facilities. Under this scheme, each school is provided with: (i) at least two reasonably large all-weather rooms along with separate toilet facilities for boys and girls; (ii) at least two teachers (one male and one female); and (iii) essential teaching and learning materials including blackboards, maps, charts, a small library, toys and games, and some equipment for work experience.

External evaluation of the scheme indicated the lack of training of teachers in using the teaching materials, specification of a large number of uniform facilities to be provided without modification according to local needs, and lack of provision for breakage of equipment. Effective steps have since been taken to remove these drawbacks. The scheme of Operation Blackboard has also been modified and expanded to provide a third room and a third teacher to primary schools where enrolment exceeds 100, and it has been extended to upper primary schools. The scheme is concentrating on rural areas and SC/ST areas, and girls' schools are being given the first priority.

Establishing Linkages between Programmes of early Childhood Care and Education (ECCE), Primary Education, Literacy and UEE

Early childhood care and education (ECCE) is viewed as a crucial input in the strategy of human resource development, as a feeder and support programme for primary education, and as a support service for working women of the disadvantaged sections of society. Since the age-span covered by ECCE is from conception to 6 years, emphasis has been given to a child-centred approach and play-way and activity-based learning in place of formal methods of teaching including introduction of the 3 Rs. Keeping in mind the role of ECCE as a support service in UEE, it is deliberately directed to the most underprivileged groups, those who are still outside the mainstream of formal education. The aim of ECCE is that every child should be assured access to the fulfilment of all basic needs. It involves the total development of the child in every aspect including the physical, psychomotor, cognitive, language, emotional, social and moral. The present ECCE programmes include:

- the integrated child development service (ICDS)
- the scheme of assistance to voluntary organisations for running early child education (ECE) centres
- *balwadis* and day-care centres run by voluntary agencies with government assistance
- pre-primary schools run by state governments, municipal corporations and other agencies

- maternal and child health services through primary health centres, sub-centres and other agencies

The ICDS is today the biggest programme of early childhood development, serving about 15 million children and 3 million mothers.

Appropriate linkages are being established between ECCE programmes, primary schools, NFE centres and other related schemes of UEE.

Promotion of Access to Girls and Disadvantaged Groups

As with all educational indicators, gender disparities are conspicuous in regard to enrolment and retention. Over the past 25 years, enrolment of girls at the primary stage has grown from 5 million to 47 million and at the upper primary stage, from 0.5 million to 16 million. But disparities persist. Today girls account for only 46 per cent of the enrolment at the primary stage and 38 per cent at the upper primary stage. The drop-out rates of girls at the primary and upper primary stages are higher than those of boys. Regional disparities are also conspicuous. The very low female literacy (20 to 29 per cent) in some of the major north Indian states causes grave concern. The rural girls are doubly disadvantaged by non-availability of educational facilities and by their domestic chores.

Concerted efforts are now on to reach out to the girl child in rural and remote areas and urban slums by designing special NFE programmes with a view to getting them back into the formal stream. The NFE programmes are being dovetailed into the total literacy campaigns (TLC) to reach out to the girls in the 10-20 age group. Programmes for continuing education are being designed to ensure that neo-literates and school-going girls have access to reading materials.

An important constraining factor for female education is the lack of women teachers in rural areas. Therefore, special efforts are being made to recruit women teachers and to augment teacher training facilities for women so that adequate numbers of qualified women teachers are available. Co-ordinated efforts are also on to provide the necessary support services to enhance their participation and performance.

We in India are unambiguous about removal of disparities and attainment of equality of education opportunities for SCs, STs and other backward sections including girls. A number of strategies aimed at accelerating their rate of enrolment and retention have been detailed and are being implemented. Because of the affirmative policies of the government, the enrolment of these categories has increased considerably at the primary stage. The participation of SCs and STs at the primary level is more or less in proportion to their share in the population. Drop-outs, though declining, continue to be significantly large [primary stage (classes I-V), SC 49 per cent, ST 64 per cent; upper primary stage (classes VI-VIII), SC 68 per cent, ST 79 per cent]. Gender disparities are conspicuous among SCs and STs.

To ensure universal access and enrolment of SC children in rural areas, priority is given to the needs of SC habitations and hamlets in opening primary and upper primary schools. For SC children access and enrolment are assured primarily in the formal schools. Where they are not able to attend these, provision is made for non-formal and distance education centres. Every ST habitation is being provided with a primary school or other suitable institution. In tribal areas, the educational plan is being implemented in an integrated manner. Pre-school education, non-formal education, elementary education and adult education are being organically linked and integrated to ensure achievement of total literacy of the entire population.

Adequate incentives are given to the children of SC, ST and other backward sections in the form of scholarships, uniforms, textbooks, stationery and midday meals. All schools, NFE centres, and pre-school centres in SC/ST habitations are being equipped with necessary infrastructural facilities in accordance with the norms laid down for Operation Blackboard and for achieving MLL. Operation Blackboard has already covered almost all schools in tribal areas. Indigent SC/ST families are given incentives to send their children, particularly girls, to school.

Restructuring of Teacher Training

Teacher performance is the most crucial input in the field of education. In the ultimate analysis, the national policies on education have to be interpreted and implemented by teachers as much through their personal example as through teaching-learning processes. With a view to improving the quality and competence of teachers, a centrally sponsored scheme of Restructuring and Reorganisation of Teacher Education (RRTE) was launched in 1987.

During the period 1987-90, nearly 1.8 million teachers were trained under the programme of mass orientation of school teachers (PMOST). Most of them were primary and upper primary teachers. The main objective of the programme was to orient teachers in the main priorities and directions envisaged in the NPE 1986 and to improve their professional competence.

Among the other main components of the RRTE, as far as elementary education is concerned, are:

1. setting up of District Institutes of Education and Training (DIETs) in all districts to provide good quality pre-service and in-service training to elementary school teachers and adult education/non-formal education personnel and to provide resource support to these systems
2. organising Special Orientation Programmes for Primary Teachers (SOPT) with a view to providing training to teachers in the use of OB materials and orienting them towards MLL strategy with a focus on teaching of language, mathematics and environmental studies

More than 300 DIETs have already become operational and have started conducting training programmes. The SOPT launched in 1993-94 is now going on in almost all states and more than 115,000 teachers have already been trained. A National Council for Teacher Education (NCTE) was set up in 1993 with statutory status for the effective implementation of all teacher education and training programmes and to achieve planned and co-ordinated development of the entire teacher education system throughout the country. The regulation and proper maintenance of norms and standards in the teacher education system is the responsibility of the NCTE.

Availing of External Financial Support for Basic Education

As a matter of policy and principle, India had not been seeking financial support from external agencies to implement its programmes of basic education. This situation changed in 1991-92, when a conscious and strategic decision was taken to avail of external assistance to achieve the goal of Education for All (EFA).

Today a number of agencies including the World Bank, UNESCO, UNICEF, Swedish International Development Agency (SIDA), International Development Association (IDA), and the British Overseas Development Agency (ODA) are sharing our concerns in this area. A new phase has, therefore, emerged — a phase of partnership between the inherent potential of the country and financial and other support from external agencies.

Launching the National Elementary Education Mission (NEEM)

With the objective of mobilising all the resources — human, financial and institutional — necessary for achieving the goal of UEE by the year 2000, a National Elementary Education Mission (NEEM) was set up in August 1995 with the District Primary Education Programme (DPEP) as its core. This Mission will monitor and implement all the meticulously formulated strategies based on microplanning, and will ensure that free and compulsory education of satisfactory quality is provided to all children up to 14 years of age by the turn of the century.

15

The School System in India

A critique*

Neeru Nanda

THE educational system in India has faced a basic dilemma ever since its introduction by the British. The essence of this problem was summed up by Mahatma Gandhi in his historic statement at Chatham House, London, in 1931.

The British administrators when they came to India, instead of taking hold of things as they were, began to root them out. They scratched the soil and began to look at the root and left the root like that and the beautiful tree perished. The village schools were not good enough for the British administrator so he came out with his own programme. Every school

---

* This article is by way of a critique of the system of education introduced in India by the Briush and further expanded and developed by the Indian government. India today has a resource pool of skilled manpower which can compete with the best in the world. The present paper, however, has not been written from the point of view of those who have benefitted from this system.

Mahatma Gandhi had once prescribed an acid test for judging the efficacy of any scheme by visualising the poorest citizen of this country and then analysing how that programme could benefit him. This paper has accordingly been written from the point of view of those who have failed to benefit from the system. It has been written on behalf of the illiterate, the underprivileged, the drop-outs, as also those who are struggling to join the ranks of the privileged. It has been written on the basis of assumptions spelled out in the preamble prepared for this conference. In brief, that education cannot be limited to technical knowledge or skill. It is essentially a practice of prudence concerned with justice and humility, a pursuit of living in an interconnected universe.

The author is solely responsible for the views expressed.

must have so much paraphernalia, building and so forth. Well, there were no such schools at all — ancient schools have gone by the board — and the schools established after the European pattern were too expensive to fulfil a programme of compulsory primary education of these inside of a century. This very poor country of mine is ill able to sustain such an expensive method of education.

Free India did not have the will to fulfil Mahatma Gandhi's dream of reviving the ancient tradition of the village schoolmaster, supported by the community, his status and survival as a *guru* being embedded within the culture and ecology of his immediate environment. Instead the government chose to continue with efforts to educate the masses through a vast, centralised machinery and superstructure of staff, infrastructure and resources. Successive efforts at universalisation of primary education, ranging from Operation Blackboard to DPEP (now funded by the World Bank and other international agencies) and Education For All, have only reinforced the strength of the challenge posed by Mahatma Gandhi in his statement at Chatham House. The Adult Literacy Campaign and, more recently, the investment thrust on primary education to produce the **literate child**, carries within itself the hidden suspicion that the goal of universal education for the masses may even be quietly set aside one day, planners having perhaps finally realised that the goal is unattainable within the present system.

As pointed out by the Mahatma, the success of this system depends solely on expenses and affordability. It thus gives rise to a situation where a mass of underprivileged and under-educated children face a largely urban elite population of privileged schoolchildren. Needless to say, it is the latter who get into the universities. While the tax-payer pays for both government schools as well as universities, the poor man pays for both but benefits from neither. The per capita public cost of university education is exorbitant when compared with school education. In the words of Ivan Illich, School is a perfect system of regressive taxation where the privileged graduates ride on the back of the entire paying public.

Every poor Indian who can manage to make both ends meet thus strives to join the ranks of far better (and more expensive) schools, because only these can form the gateway to subsidised higher education. The past two decades have seen a phenomenal growth of two types of private schools outside the system of government schools (which have also, incidentally, grown by leaps and bounds, despite resource constraints). These are the so-called elite schools where education is sold to the highest bidder, and the plethora of petty teaching shops against which it is fashionable for the educated elite to occasionally raise a hue and cry. In between there do exist a large number of institutions, often run by charitable or religious organisations, which seek to make as many compromises as possible within the given system.

It is not surprising that the government of a democratic state would also be compelled to join this race for providing more expensive schools for quality education. The Kendriya Vidyalayas, Sainik Schools, Railway Schools, Tibetan Schools and (more recently) Navodaya Vidyalayas are responses to the basic dilemma facing the existing system. The KVs and NVs together run about 1,200 schools. The KVs, which were set up to cater to children of government servants having All-India transfer liability, have now increasingly opened their doors to the children of upwardly mobile, less privileged urban families. While 'political pressure' is ostensibly to blame, the trend basically reflects a sensible popular response of the underprivileged. In the public perception, the system of mass government-sponsored education appears to have failed to deliver the goods, increasing expansion having led apparently to decreasing quality. Hence the upwardly mobile social classes are aggressively latching on to whatever better education is available at public cost.

As regards rural India, it needed a statesman like Rajiv Gandhi to visualise that the day was not far off when the rural elite would also grow restive and clamour for expensive quality education at par with quality schools set up in urban areas serving predominantly urban populations. Hence Navodaya Vidyalayas were set up as residential schools with 75 per cent reservations for rural candidates, all candidates being selected on the basis of a nation-wide talent test. No doubt the Navodaya system has made the best possible compromises within the given contradictions of the existing social system. There is a 30 per cent reservation for girls and reservation for scheduled castes and scheduled tribes commensurate with their population strength in the district. Elitism of merit is the hallmark of the system. While the talent test is open to all rural children studying in class V of rural schools, the admissions are also structured so as to ensure that all geo-economic regions within the district are proportionately and equitably represented and no single region dominates.

Navodaya Vidyalayas have sometimes been sarcastically described as the Doon School alternative for rural areas. But this can mean different things to different people. From the urban point of view it is often a derisive comment implying that by the very nature of the socio-economic set-up of the country, Doon Schools can hardly be relevant for country bumpkins. For the rural elite it is viewed as an optimistic statement, since they would naturally like to have the best for themselves, the urban elite invariably furnishing the role model.

There is a third point of view put forth by planners and educationists who feel that these rural elite schools are being set up at the cost of improvements in mass universal education. Such expensive schools set up in the rural environment only bring into focus the deprivation suffered by the rural masses. Setting aside expenditure on construction, which is quite heavy (each school complex costing over Rs.4 crores), the average cost per Navodaya student (including all administrative and other expenses) comes to about Rs. 9,900 per annum. Is such heavy investment on as few as 500-odd students per district justified? These are the questions raised by the Acharya Ramamurti Report, as also by the CPI(M) government of West Bengal, which has steadfastly refused to introduce the Navodaya Vidyalaya Scheme in its state purely for ideological reasons.

The Acharya Ramamurti Report indicts the NV Scheme on the grounds that it is very costly, it caters to a microscopic minority of the total school population and perpetrates an exclusive system inconsistent with the 'long cherished common school system of public education'. These educationists, however, seem to have overlooked the main problem pointed out by Mahatma Gandhi with the 'cherished system of public education', namely, that if at all schools are to be established after the European pattern they will inevitably be too expensive for the people. If rural areas are not to follow the pattern of more and more exclusive and expensive schooling mushrooming in urban areas, rural talent will never get a chance to compete for higher education unless it migrates to the urban environment. This is a hard fact of life which this point of view does not face but which could be realised. By initiating the Navodaya Vidyalaya Scheme, government simply brought the process of handling contradictions inherent within the system to its logical climax. To the urban elite, however, what is sauce for the goose is definitely not sauce for the gander.

In the trend towards exclusive and increasingly expensive education the residential school stands at the apex of the school system and is sought to be duplicated for educationally underprivileged and rural children by the Navodaya Vidyalayas. As regards day schools, it is no secret that home tuition is embedded in the system of even the best public schools today. Both schools as well as parents find themselves unable to cope with the existing syllabus as well as an increasingly competitive environment without the aid of home tuitions, which today comprise a flourishing business. In government schools the underprivileged may also suffer since the temptation to earn more as a side business tends to retard optimum contribution of government teachers within school.

Thus even the expensive day school has ended up in total dependence on a parallel and equally expensive system of home coaching. A recent development in the private sector has now led to the opening of even more exclusive elite schools which restrict the number of students per class. While sixty students are replaced by twenty students only in these super schools, the costs of sixty are naturally paid for by the twenty privileged families. School has, therefore, become a flourishing and money-guzzling industry. Children exist for schools, not schools for children, and this is as true for private paying schools as it is for government schools.

While the seeds of spiralling expenses have grown into monstrous weeds, the seeds of alienation spread in a more subtle manner. The residential school is an apt symbol of alienation not only from its cultural and social environment but also from society itself. It stands at the apex of a system which rests on this artificial separation that seeks to inculcate values of exclusivity and elitism. Within the day schools the system alienates the teacher from the taught through the weight of sheer numbers and the cumulative burden of learning and measuring up to academic standards. It also alienates the student community as a whole from the socio-cultural matrix which surrounds it. The accent on competition and the constant stress on the individual as against the group and the community further alienates children from one another. Seeds of alienation are embedded in the examination system, which lays a premium on a system of marking that is the sole determinant of individual achievement, the costs of which are paid for by the under-achievers. In its mildest form the under-achievers pay for it by being relegated to 'non-ability' as against 'ability' sections in public schools, alienation from one's immediate peers being thus glorified and institutionalised. At its worst the system is scarred by student suicides.

Universities like Taxila and Nalanda flourished and attained international acclaim. In the ancient systems of learning the *guru* invariably utilised the older boys to teach the younger ones. Their roles were supportive, not exploitative. How unlike the system of public schools, where the exploitation of younger boys by older boys is dubbed respectably as 'ragging'. While the school tie and similar indicators of superior status may still hold them together at school, these tenuous bonds no longer operate in the universities. University students in India in particular are in essence a mob, whether it be a passive mob or a mob on the rampage.

There are often deaths of students in which their peers are indirectly or directly involved through ragging. It is not a mere coincidence that these incidents have generally occurred in professional institutions for engineering or medical students. While at first glance this phenomenon may appear to be incomprehensible, once the seeds of alienation present in the whole system are exposed it could also be seen as an offshoot of the system itself. The idea and essence of peace and universal harmony, of oneness with nature and association with the earth, of education as a pursuit in the arts of living in an interconnected universe is entirely missing from the system. Instead to the fore are values of individual achievement, branding and typing young persons as bright, mediocre, elite, poor, underprivileged. Divisive forces predominate over harmony. Products dominate over people. Competition dominates over brotherhood.

That this system is giving rise to problems entirely of our own making is being realised in many parts of the world. Radical thinkers, sociologists and educationists have raised voices in protest. Everett Reimer in *School Is Dead* and Ivan Illich in *Deschooling Society* have asked some of the profoundest questions about education today. They have recognised that most countries in the world can only afford to give their children the barest minimum of education while the costs of schooling are rising everywhere faster than enrolments and faster than national incomes. While regarding schools as institutional props for privilege for most people, Everett Reimer also recognises the fact that they are at the same time major instruments of social mobility. But he questions the cost in terms of true learning, true creativity and true democracy.

The search for alternatives in education has already begun. Some, like Everett Reimer, search for alternatives in terms of content, organisation and finance. At the other end of the spectrum, *Letter to a Teacher*, written by eight young Italian village boys from the school of Barbiana, is a minor classic which reveals with devastating clarity how the school system oppresses and marginalises the rural poor by divorcing them completely from their ecological base and their cultural environment. It also presents a radical alternative.

In India today the government's and NGOs' search for alternatives has led to greater support and funding for non-formal schools, the most dramatic example being that of the Charvaha Schools of Bihar. However, this experiment suffers from all the general weaknesses of the state system. In Bihar there are other alternatives developed with more success, such **as the model** 'Sidh', started by Pawan Kumar Gupta, an NGO, in and around about 15 villages surrounding Kempty, near Mussoorie. Many other examples of rural-based education centres can be documented where the initiative for establishment and management has vested with communities themselves. Craft-oriented training and basic education have been outstanding successes of Dwarko Sundrani's Bodh Gaya Ashram, where schooling on the Gandhian pattern is made available to the poorest of the poor in Bihar. These non-formal schools, being beyond the pale of the state boards, are also free from the usual bureaucratic supervision of school inspectors, but by their very nature they are limited to the class V level.

The Yashpal Committee was set up specifically for suggesting ways to reduce the academic burden on students in recognised schools. The Committee strongly recommended toning down of individual competitiveness and introduction of group activity in schools. Simultaneously it advocated decentralisation of curriculum formation, textbook writing and much greater involvement of teachers in these processes, particularly textbook formation. In general the Yashpal Committee came out strongly against textbooks, syllabi and examinations which together form a system that inflicts upon the children the 'tyranny of rote memorisation.

The Department of Education has taken up these suggestions for serious review. The CBSE pattern of education will be set aside or modified in the near future. Sir Thomas

Munro, who set in motion the process of establishment of schools after a new pattern in Madras Presidency in the 1820s, aptly summarised the aims of the new education as being 'to facilitate the acquisition of wealth or rank'. As long as society accepts these as the aims of schooling, the CBSE syllabus and the emphasis on science and commerce streams (putting incalculable strain on the students) will continue regardless of the Yashpal Committee report or any other similar reports. There are a few notable public schools (like Rishi Valley of the J. Krishnamurti Foundation, Banasthali Vidyapeeth of Rajasthan, the Netarhat residential school experiment of Bihar) which have departed from the mainstream and yet retained eminence within the general delivery system.

The difficulties of innovating within over three hundred and seventy government residential schools such as the Navodaya Vidyalayas would be apparent to any educational administrator. Being wholly funded by government, there is an even greater compulsion for these schools to deliver and perform as per the marks-oriented examination system laid down by the CBSE. Not only do residential schools allow for more investment in time, but the Vidyalayas themselves, with their resource pool of talented rural students, stand Janus-like between two worlds — with one hand they can reach out to grasp the best of scientific and technical skills, while with the other they can gather easily the ancient wisdom of the east enshrined in the oral tradition and rituals of their immediate cultural environment. Hence the Samiti has started a small venture to give schooling a human face through a carefully planned programme of Art in Education, based firmly on the indigenous *guru-shishya* tradition of learning.

Workshops in theatre, folk music, folk arts, epics and ballad singing can very easily slip into the groove of consumerism and assembly-line production that has come to characterise our school system. Again, overemphasis on achievement and the accent on the individual could vitiate the cathartic experience, the flowering of creativity, the realisation of a deep harmony between man and nature which these workshops should generate. Hence every workshop in Navodaya Vidyalayas is designed around an activity with a cluster of groups working on the theme in accordance with their own particular talent. Thus a Pandawani workshop held in Durg was not just an exercise to locate another Teejan Bai amongst the students and project her on the stage. It was an experience in which the literary group interviewed Teejan Bai, wrote up reports and poems, while the visual art group was busy drawing and painting the scenes she depicted. Those who had talent for dramatic performances learnt the art and performed on the stage, but it was a holistic experience through which the Chhattisgarhi children of Durg moved towards a joyous self-attainment, a recognition of an abiding and ancient identity which was derived from sources that stand apart from and are superior to their school and their schoolmasters. The richness of rural roots and traditions gave them a sense of pride and psychological reassurance which no success in examinations could ever give. By honouring notable rural artists in this fashion the Samiti was able to bridge the gulf between teacher and taught, urban and rural, so-called literate and so-called illiterate. To put it in the works of Dharampal, the present system "has kept most educated Indians not only ignorant of the society they live in, the culture which sustains this society and their fellow beings, but yet more tragically, for over a century it has induced a lack of confidence and loss of bearing amongst people of India in general". It is a tribute to the children that they could very easily grasp the essence of the programme best summed up in a poem written by one girl who exclaimed "Aha! Look at her! She has not been to school even for a single day and yet what greatness she has achieved. Look at us, who go to school everyday, it is for us to get inspired and learn from her!"

It is ultimately left to each one of us in the educational field to move one small step forward towards this transformation of vision towards reorientation of values and objectives of education. Not only in India, bu+ in the entire world, education places an overriding value on products over people, on achievement over harmony. Ivan Illich has expressed this need in moving terms, quoting from Yevtushenko's poems.

> We now need a name for those who love people more than products,
> those who believe that
> No people are uninteresting
> Their fate is like the chronicle of planets.
> Nothing in them is not particular and planet is dissimilar from planet.
> We need a name for those who love the earth on which each can meet the other.
> And if a man lived in obscurity
> making his friends in that obscurity
> obscurity is not uninteresting.

16

Innovations in Primary Education in Bangladesh

A.M. Sharafuddin

BANGLADESH is generally known for its high population density (over 800 persons per square kilometre), high population growth rate (1.8 per cent per annum), endemic poverty (per capita GNP about $ 240) and high rate of adult illiteracy (58 per cent). However, in recent years some innovative programmes of basic education have been initiated in the country to tackle the overwhelming problem of illiteracy.

Agriculture is the mainstay of the economy of Bangladesh, and 80 per cent of around 120 million people live in 86,000 villages. The population is basically rural, but urbanisation is growing at the rate of 5 per cent per annum, which is almost three times the population growth rate. The main reason for the high rate of urbanisation is the shifting of the rural population to the cities in search of jobs, creating a large number of slums in urban centres.

The impact of the Jomtien World Conference on Education for All (WCEFA) held in March 1990, which promoted an expanded vision of basic education with the aim of meeting the basic learning needs of all children, youths and adults, has been felt in Bangladesh. As a result, a Compulsory Primary Education Act was passed in 1990 and, in January 1992, the government launched the Compulsory Primary Education (CPE) programme in 68 out of 460 *thanas* in the country. This programme has been extended nation-wide since January 1993. About the same time, in 1992, the government made rural primary education free for girls up to grade 8, and a new Primary and Mass Education Division (PMED) was created under the Prime Minister's Secretariat to strengthen the management of primary education and non-formal education programmes.

The national EFA programme was officially launched in March 1992. A project called

Expansion of Integrated Non-formal Education Programme (INFEP) was also initiated; this has recently been upgraded into the Directorate of Non-formal Education. In 1993 the government launched the 'Food and Education' programme, designed to improve enrolment and class attendance and reduce drop-outs. For qualitative improvement in the primary education sector, several new programmes have been introduced such as (a) a curriculum dissemination programme for teachers, (b) cluster and sub-cluster training programmes, (c) continuous pupil assessment, and (d) a satellite school programme.

Non-formal Education Programmes

In the 1980s a number of NGOs began to see the need for initiating programmes of basic education in support of their poverty alleviation activities. An important result of the Jomtien WCEFA has been that several NGOs have identified education as a priority sector for their areas of operation. Many of them have significantly expanded their education programmes over the last few years.

The NGOs realised that educating the whole nation would require a massive effort and close collaboration between them and the other progressive forces. Consequently, in 1990 a coalition of the mainstream NGOs having major education programmes was formed under the title 'Gono Shakkharata Ovijan' (Campaign for Popular Education — CAMPE). The aim was to foster cooperation among the NGOs, the government and civil society in developing a mass movement encompassing both formal and non-formal programmes in order to eradicate illiteracy.

Currently more than 400 NGOs are involved in non-formal education programmes in Bangladesh. In 1994, these had more than 2.6 million enrolled in about 73,000 centres. Of those enrolled, at least 1.4 million were primary-age children (63 per cent of them girls), over 0.4 million were adolescents (65 per cent girls), and about 0.8 million were adults (80 per cent women).

In the EFA National Plan of Action, the government set the target of raising gross primary enrolment from 76 per cent in 1991 to 95 per cent by the year 2000, the completion rate from 40 per cent to 70 per cent, and the adult literacy rate from 35 per cent to 62 per cent. It is being claimed that already considerable progress has been achieved in realising these targets: gross primary enrolment has reached 92 per cent, completion rate 60 per cent and adult literacy rate 42 per cent.

Innovations in Primary Education

In view of the large number of children and adults who remain uncovered by present educational programmes, it is felt that some innovations are needed in primary education to provide quality education at low cost which would ensure high attendance and retention as well as a high rate of success. Several innovative programmes have been initiated in primary education by both governmental and non-governmental agencies to achieve these aims. Generally, the major characteristics of an innovative programme are that it should (a) be based on direct experience and practice; (b) be cost effective; (c) show easily verifiable results; (d) have interdependence within its subsystems; (e) have in-built mechanisms for measuring achievement; (f) have simple but effective management; and (g) create a strong team spirit among the members of the programme. Based on these criteria, some of the more prominent and successful innovative programmes in the field of primary education in Bangladesh are as follows:

1. *BRAC Non-Formal Primary Education Programme*: The Bangladesh Rural Advancement Committee (BRAC) has developed this programme. This caters to children 8-10 years of age who could not get admission in formal primary schools. Thus the learners are older children who never attended school. The programme covers grades 1-3.

2. *GSS Primary Education Programme*: This programme has been initiated by the Gono Shahajjo Shangstha (GSS). Children who could not get admission in formal primary schools are enrolled in these schools. The children are taught up to grade 3, which is now being extended to grade 5. Classes are generally held in brick schoolhouses built by GSS.

3. *CMES Technology School*: The Centre for Mass Education in Science (CMES), a science- and technology-oriented NGO, has initiated this programme. In grades 1-2, pupils are provided general education; in grades 4-5, some science lessons are included covering environment, popular science and technology. The aim is to make people conscious of the importance and impact of science education in real life.

4. *Government Satellite School Programme*: Under this programme, 'feeder schools' are set up within the catchment areas of the formal primary schools so that small out-of-school children who are unable to walk a long distance can receive early education in grades 1-2. The aim is to bring school facilities near the homes of poor children, especially girls, to ensure school attendance. Teaching is done by a couple of volunteer women teachers who get a small honorarium. The schools are run entirely by the local communities.

5. *Dhaka Ahsania Mission Alternative Primary School Programme*: Children of 6-8

years are the target group. The course duration is 27 months with 3 grades, each of 9 months. The school premises are provided by the local community.

6. *UCEP School:* The Under-privileged Children's Programme School is a specialised programme designed for working children in the age group 6-14 years. The duration of this programme is 7 years. It is a mixture of general and technical education.

7. *Terre Des Hommes (TDH) Street Children Programme:* Children aged 8-10 years who work at railway stations, bus and launch terminals, etc., are the target groups of this programme. Such children are assembled for teaching under the sheds of railway stations or bus and launch terminals with the help of local authorities including the police. BRAC curriculum and materials are used.

Many of these programmes have shown that even with very limited facilities, through better design of materials, improved teaching methods, and better management and supervision, schools can be made highly attractive to children so that drop-outs and absenteeism are almost totally eliminated. The details of some of these programmes are given here.

Models of Non-Formal Primary Education

(A) THE BRAC MODEL: NATION-WIDE NFPE COVERAGE

BRAC was one of the first NGOs to have started large-scale programmes of non-formal primary education in Bangladesh. While BRAC started as a relief and development agency for the rural poor in 1973, its education programme began with only 22 experimental schools in 1985. Today its NFPE programme is by far the largest single non-government primary education programme in Bangladesh, currently with more than 30,000 schools and about a million pupils. More than 90 per cent of the children who start in BRAC schools graduate, and a large proportion of the NFPE programme graduates are admitted into grade 4 or higher of the government school system.

BRAC organises two types of schools: three-year NFPE schools for 8-10-year-olds who have never attended school; and two-year Kishor-Kishori (KK) schools for 11-16-year-olds who have dropped out of primary school and are unlikely to return. The major components of the NFPE programme are the following:

Pupils: A school consists of 33 children, 72 per cent of whom are girls living in rural areas, within about a two-kilometre radius of the school. For the most part, pupils come from relatively disadvantaged homes, their families being landless or owning only their homesteads.

Teachers: Teachers are generally married adults, 95 per cent women, who have completed nine or more years of education and live within easy walking distance of the school. The teachers are hired on a temporary, part-time basis and are paid modest wages. There is one teacher for every 33 pupils. Teacher training includes 12 days of initial training at a residential BRAC training centre and one-day or two-day refresher training sessions each month conducted by BRAC staff at a BRAC office near the teacher's school. Weekly visits from BRAC field workers provide regular feedback.

Parents: The parents of most BRAC school pupils are illiterate and are usually the most socio-economically disadvantaged in their villages. Parents pay no fees for the schooling, apart from replacing broken slateboards and worn mats; BRAC provides all pupil and teacher supplies—pencils, notebooks, textbooks, teacher manuals, slateboards, chalk, etc. Prior to the opening of a new school, parents and BRAC staff meet several times. Parents must pledge to attend monthly meetings and to send their children to school each day.

Schedule: The NFPE instructional programme is presented in three-year cycles. The school is in session for 2½ to 3 hours a day, six days per week, 268 days per year, at a time of day selected by the parents. The group of 33 pupils is enrolled at the beginning of the programme and advances together through the programme. At the end of the programme, the school begins another three-year cycle if there are enough eligible children in the community.

Instructional site: Instruction is provided in one-room premises rented for just three hours per day. These rooms generally have bamboo and mud walls, an earthen floor, a tin roof, and a blackboard. The children sit on the floor on bamboo mats, holding their slateboards on their knees. The teacher has a stool and a metal trunk that serves as a desk as well as a supply cabinet.

Curriculum: The curriculum for both NFPE and KK schools, consisting of Bangla, social science and mathematics, has developed over a period of years and has been revised several times. The material covered is roughly equivalent to grades 1-3 in the formal school system. Since the formal school system requires English, the NFPE schools include English in their curriculum during the third year so that children who want to join formal schools later are well prepared.

BRAC's ability to implement all the necessary elements of a targeted basic education programme derives largely from its expertise in development management. In terms of development, since its creation 20 years ago BRAC has designed surveys that help it develop and target its programmes for the most disadvantaged rural families. BRAC also has developed ways to encourage these families to participate in the decisions that most affect programme implementation.

Cost efficiency: Independent cost studies have confirmed BRAC costs for schooling (about Tk.800 or US $20 per annum) as roughly equal to the government's formal schooling, without considering the extra private costs that make the formal schools more expensive and lead to high drop-out and low enrolment rates in the formal schools. In addition, unlike the formal school system, which allocates most of its resources to teachers' salaries and school facilities, BRAC allocates almost 30 per cent of the NFPE programme budget to management and supervision. Only 29 per cent is allocated to salaries and 6 per cent to rent school space.

BRAC pupils achieve as much as or more than formal school pupils. BRAC pupils complete the NFPE programme and enter the formal grade 4 at a higher rate than do formal school pupils. BRAC pupils score as much as or better than formal school pupils in basic education assessment and basic literacy tests.

Even when annual costs per enrolled pupil in BRAC and the formal school system are approximately equal, the relatively higher attendance rates, lower repetition rates, higher grade 3 completion rates, and higher grade 4 continuation rates for BRAC pupils mean that BRAC schools are substantially more cost efficient per graduate than the government's formal schools.

Instructional schedule: BRAC schools meet for 2½ hours per day for 268 days in a year. School is held six days a week and the school hours are flexible, depending on the convenience of the parents. The school schedule allows for a short vacation, which is determined jointly by parents and teachers. Teacher absences are quite low. Because of the relatively low pupil-teacher ratio, teachers require little time to take roll and no time is wasted moving from one grade to another. In addition, BRAC teachers assign little homework and consequently spend a minimum of class time on it.

By contrast, government schools operate 220 days a year for one or two hours, two or three hours shifts per day. In addition, teachers may have as many as 100 pupils in a classroom, at least tripling the amount of time allocated to simply taking roll. Because most of BRAC's classroom time is allocated to instruction, the estimated annual range of actual instruction is between 670 and 804 hours.

Instructional site: BRAC schools operate in rented one-room premises at least 240 square feet in size. BRAC rents these buildings for less than US$5 per month. Pupils sit on woven mats on a mud floor in a 'U' shape, with a blackboard and teaching aids at the front of the classroom. Neither the teacher nor the pupils have desks. All books and stationery are supplied by BRAC.

The distance from home to school for BRAC pupils ranges from less than 1 km to 2.5

km. In comparison, the average catchment area for government schools is about 3.2 km, with distances somewhat greater in remote rural areas. Because of this proximity children lose less time in travel to and from school. Especially for girls, this is considered relatively safe. Also, parents are able to monitor what happens inside the schoolroom, how their children are treated, and whether they are happy and busy.

Classroom environment: The 33 pupils that comprise a BRAC school move through three years of course-work as a group. One teacher leads the group. This pupil-teacher ratio is very low in comparison to the government primary schools, where the average pupil-teacher ratio is 65:1. Pupils are often divided into small working groups in which the quicker pupils help the slower ones and all pupils move together through the lessons at the same pace. BRAC materials stress a basically child-centred approach to learning. Instruction in the core subjects is broken up with co-curricular activities, sometimes for as little as five minutes between subjects.

Curriculum and materials: The NFPE's instructional materials have gone through several phases of revision, sometimes with assistance from the Institute for Education and Research of the University of Dhaka and several outside consultants. The curriculum originally covered three subjects: Bangla, maths and social science. By 1987 it was clear that many of the NFPE programme graduates planned to continue in government schools, and the BRAC curriculum was modified to incorporate English and religious instruction, required subjects in government schools.

The current BRAC curriculum spans grades 1-3 and includes Bangla, mathematics, social studies and English with an emphasis on the practical health and social issues that are likely to be encountered by a typical BRAC pupil. Class time is allocated in the following segments: Bangla (25 minutes reading and 25 minutes writing); mathematics (35 minutes); social studies (25 minutes); and two 20-minute co-curricular activities, which include physical exercise, field trips, singing and dancing. English is added to the schedule in grade 2.

The BRAC curriculum addresses significantly fewer objectives than does the government's primary curriculum, particularly in languages and mathematics. Such a lean curriculum may be a contributing factor to be programme's success. A curriculum that addresses fewer topics allows teachers to cover them at a deeper level than if the teacher is responsible for covering many topics. BRAC's simplified curriculum is effectively implemented, whereas the more comprehensive government curriculum is not fully implemented in most formal primary schools.

The materials are carefully sequenced, segmented into short, discrete lessons, and attractively printed in small, non-threatening booklets (as opposed to larger books that can overwhelm young learners). There is one reading primer for the eight-week preparatory phase (this pase is only two weeks for older pupils), one reading booklet and one maths booklet for each of the three primary grades, and one social studies booklet for each of grades 2 and 3. Concrete examples from everyday life are used throughout the booklets, especially in the social studies materials.

B. THE GSS MODEL: INDEPENDENT READERS AND WRITERS

GSS, a leading NGO of the country, was set up in 1983, aiming at the empowerment of the rural and urban poor. It started its activities by organising and mobilising agricultural labourers of fourteen villages of Khulna district in south-western Bangladesh. Over time it has grown into a national-level NGO which has development intervention in 16 out of the 64 districts of Bangladesh, covering over 450,000 households. GSS joined the private sector effort for promoting basic education in mid-1987 with the first education centre in a village. Since then its educational programme has been expanded to about 250 centres for both rural and urban children.

A major emphasis of the GSS is on the education of slum children. The urban population is expanding fast. It is estimated that in the capital city, Dhaka, as much as 40 per cent of the 8 million population are slum dwellers. The slums lack in such basic facilities as sewerage, drinking water, education and health care. They also have very high housing density (6-8 members in a room of 42 sq ft). According to a study by the Centre for Urban Studies of the University of Dhaka, only 9.4 per cent of the households in the slums have primary schools within their reach. Also, the existing educational system is not suitable for slum children because they have to work for their livelihood. The curriculum content has little to prepare them for life. The teaching method is also not designed to unlock the potential of these children. No wonder the rate of school attendance is low in the slum areas. Against this backdrop, GSS came forward to provide basic education to slum children.

In response to a request from groups in urban slums, GSS opened six experimental schools: four in the capital city and two in Khulna port city. These original urban schools became the testing and training ground for teaching methods and curricula. The five-year curriculum was designed to provide basic education to slum children who work in the informal sector in the urban centres. In designing the curriculum and teaching method, GSS adapted Western methods for the slum environment. The objectives of the primary education programme are:

- Preparing learners as independent readers while they are in grade 1
- Preparing learners as creative writers by achieving independence of thought and

the ability to present their own ideas in writing

- Enabling children to become numerate
- Creating a real and lasting interest in books and other printed matter
- Taking pupils to the standard of grade 5 of the formal system

It also set the following quantitative targets:

- More than 90 per cent attendance rate
- Less than 5 per cent drop-out rate for grades 1-3
- More than 50 per cent of the children going on to grade 4
- 100 per cent children should learn to use the four maths rules to 100 by the end of grade 3
- 100 per cent children should be able to do creative writing (six lines) by the end of grade 3

The learners: The children of GSS schools come from very poor families with varied occupational backgrounds, i.e., factory workers, carpenters' helpers, builders, masons, hawkers, etc. The children fall in the 4-14 year age group. Often their mothers and sisters work as maidservants and their fathers as rickshaw pullers and day labourers. The parents' income is not enough to afford food for the whole family in order to maintain a 2000 cal/day diet and to pay rent for 42 sq ft of damp space for the family. Their diet is highly deficient in protein, fat and vitamins.

Land for the school: Finding a piece of land in the city is a difficult task. GSS usually begins with a survey to identify out-of-school children of a slum and holds talks with their parents. Eventually, parents admit the necessity of education for their children and express their willingness to send their children to school if available. The GSS then holds talks with the illegitimate owner of a slum for setting up a school.

The Teacher: GSS invites applications from suitable candidates through local newspapers. The prospective candidate should have a minimum of 12 years of schooling along with a certificate from a primary education training institute. GSS prefers candidates with bachelor's degrees. For the post of Head Teacher, the minimum qualification is a master's degree in educations with 2 years' experience or a bachelor's degree in education with 5 years' experience.

Local female candidates are preferred. Most of the teachers of GSS schools are from middle-class families who live near the school. Teachers are recruited on a contractual basis, initially for one year. Upon good performance the job is renewed every year. For five schools one stand-by teacher is recruited who works during the leave of absence of a regular teacher. In special cases, a part-time teacher is also engaged.

Teacher training: Upon recruitment, teachers undergo a 3-day basic training course which is followed by 12-day initial teacher training. Besides that, monthly one-day refresher courses are conducted and a 5-day annual refresher course is held. In the training courses, teachers become acquainted with Western teaching methods and their adaptation in the circumstances of Bangladesh, especially in the slum setting. Once they start work, teachers are given constant support by the supervisor, who visits each school at least three times a week.

Curriculum and course duration: The course duration is three years for rural children (this is now being extended to five years) and five years for urban slum children. The urban schools are experimental ones where curricula and teaching methods are tested. In the rural schools, classes are held in two shifts and each school has three teachers. In designing the curriculum, GSS has tried to make the content meaningful, easy and attractive to the learners. As such, it has made provision for multiple choice of materials developed by different organisations. Notably, the GSS curriculum is largely based on the public-sector primary school curriculum. In addition to the main books, children also go through supplementary materials designed to accelerate the learning process.

Teaching method: GSS adapted some Western teaching-learning methods to the local setting. This method is substantially different compared to those practised by many other organisations in the following ways:

Comparison of Traditional and GSS Teaching Methodology

Traditional Method	*GSS Method*
1. Rote learning	Child-centred approach
2. Passive learning	Active learning; group teaching
3. Use of insufficient reading materials	Large quantity of meaningful, relevant and interesting reading materials
4. Absence of supporting materials	Use of adequate support materials
5. Inefficient use of lesson time	Efficient use of lesson time

The child-centred teaching practice followed by the GSS appears to be quite innovative in the context of Bangladesh, where formal primary schools follow a rigid curriculum and the teaching method basically remains teacher-centred. The GSS curriculum aims to provide the following skills to the learners:

(a) participating in a discussion

(b) drawing pictures

(c) writing and expressing creativity through writing

(d) reading of any materials

During the five-year course, learners at GSS schools are given a wide range of books published by the government, BRAC (the biggest provider of non-formal primary education), individual writers, and the GSS itself. As a result, children become independent readers within 8 months of joining a GSS school and independent writers in the following 8 to 12 months.

The additional materials are prepared with the context of the working children in mind, whereas government textbooks and materials tend to refer exclusively to the background and life-style of the middle class. Besides reading, writing and numeracy, the children of all the five class are involved in a wide range of co-curricular activities. The co-curricular activities include rhyme, dance, making paper flowers, playing, gardening, singing, story-telling, etc.

Teaching method: Unlike the traditional system in which there is a different teacher for each class, in the GSS system the teacher is responsible for all the different areas of curriculum in her or his class. This facilitates a close teacher-pupil relationship. GSS has adopted group teaching methods in its classrooms which allow the teacher to pay equal attention to each child and allows the children to develop according to their own ability and speed.

Classroom management: There are 30 children in each grade, with one teacher. In a classroom there are three tables: one for vernacular, one for creative writing and the third for arithmetic. Besides, there are three corners and in the middle a mat for playing with materials. At the beginning of the class all learners assemble and sit on the mat. Thereafter, eight work on the creative table, six on the vernacular table and six on the arithmetic table, two in each corner and the remaining four sit on the mat. Everyday each learner performs all the activities in turn.

The teacher moves from group to group, giving attention to individual children or to a group as a whole. Each child receives 4 minutes of individual attention daily from the teacher and spends the remaining time engaged in direct learning or such activities as doing maths exercises or purposeful activity such as reading story books or playing an educational game.

Classroom activities in grades 1-2: Classes for these two grades are held from 8 a.m. to 10:30 a.m. This period can be split into three sessions. The time allocation for the three sessions is as follows: 1st session, 8:00 a.m. to 8:35 a.m.; 2nd session, 8:35 a.m. to 9:50 a.m.; 3rd session, 9:50 a.m. to 10:30 a.m.

1st Session (35 minutes)

Activity	*Time*
Exchange of pleasantries	5 mins. (4-5 learners involved)
Story-telling	10 mins. (often continues next day)
Arithmetic	5 mins. (practical problem-solving)
Exercises in language	5 mins. (oral, using book and board)
Roll-call	5 mins.
Distribution of materials	5 mins.

Second Session (75 minutes)

On completion of the first session, teachers give tasks to learners. We have already seen how they are split up.

Third Session (40 minutes)

Arithmetic in small groups — 20 minutes. Each day the teacher arranges the learners on the mat for arithmetic. Learners are organised in three groups and each consists of four.

Schoolhouse: Unlike many other NGOs, GSS constructs semi-pucca buildings for the school which will eventually be retained by the community. Land is donated by the community and the construction cost is borne by GSS.

Classroom management: The classroom has a range of furniture and this is placed in a corner of the classroom. There is a mat on the floor of the classroom. The learners assemble on the mat at the beginning of a class. On entering the classroom the teacher asks the children to spread it on the floor. Learners use the mat during roll-call and story-telling.

The teacher sits on a stool in a corner where learners can have a clear view of her. This seating arrangement is believed to strengthen the teacher-learner relationship. A board is hung in a suitable corner. There are three tables in a classroom, which are used to carry out group activities. The tables are placed in such a way that the teacher can go around them to see the learners' group work. There is a bookshelf which contains books, exercise books, pencils, etc.

Wall charts are hung in the classroom. The teacher decides the topics for them. Usually they are prepared on animals, birds, flowers, fruits, vegetables, names of the months, names of the days, names of the seasons; and maps of the district, country, and the world are also made. Commonly used words are also written on them. A wall chart is usually hung at the height of the children's heads and written or drawn in coloured ink. The wall chart changes every month.

There are also wall paintings as a part of the weekly projects. Learners draw pictures and put down their own ideas and observations on paper. The grade 1 learners describe to the teacher what the drawing represents. The teacher then writes down the child's words. Children draw pictures for the wall paintings every week. At the beginning of the following week, the teacher hangs the painting so that upon arrival at school, children can see their work on the wall.

Classroom activities: GSS runs urban schools in three shifts. The first and third shifts are usually for the learners of the play groups, grades 1 and 2. The second shift is for all grades together. Here is a description of the activities of the different grades.

Grades 1 and 2: Classes of these two grades are usually held from 8:00 a.m. to 10:30 a.m. The total instruction period is 2½ hours. This time is divided into three parts as follows: first part, 8:00 a.m. to 8:35 a.m.; second part, 8:35 a.m. to 9:50 a.m.; third part, 9:50 a.m. to 10:30 a.m. The first 35 minutes are devoted to activities in which the whole class participates. The following structured tasks are conducted by the teacher:

(a) Story-telling	10 mins. (if not finished, will continue next day)
(b) News	5 mins. (4 or 5 pupils give news to the class)
(c) Arithmetic	5 mins. (practice on board and using learning aids such as sticks)
(d) Language	5 mins. (oral, board and book)
(e) Roll-call	5 mins.
(f) Materials distribution and group singing	5 mins.

Once the learners are involved in their respective tasks, the teacher goes to table 1 along with the reading record book, the reader, and a pencil. She asks a learner to read with her. She first pronounces a word from a sentence and then the learner follows her. In this way, every learner reads a sentence, in which 1½ minutes of time is spent. The reading exercise of 30 learners takes 45 minutes.

For arithmetic, creative activity and games, the teacher spends one minute on each child. For each task a total of 30 minutes are spent. Thus all the activities are completed within 75 minutes.

This arrangement creates an intense learning environment. Seated in small groups in the classroom, pupils attend to their assignments with concentration. One group solves maths problems, with each child at his or her own speed and level, another group practises writing skills in the Bangla copy-books, while a third group works with varied individual assignments. The fourth group plays with learning materials.

The time allocated for the third part is 45 minutes. Two group tasks are also done in the third part:

(a) *Reading game (20 minutes)*: Four to five groups are formed, based on the attendance of learners, each group consisting of a maximum of seven learners. Each group practises reading seated on a mat for 5 minutes.

(b) *Mathematics (20 minutes)*: Three groups are formed and each group consists of 5 learners. These groups also sit on the mat and practise mathematics for 6 minutes.

Grades 3, 4 and 5: Learning activities for the 3rd, 4th and 5th grades usually start at 11:00 a.m. and continue until 2:00 p.m., a total of three hours. Activities are divided into four parts as follows:

First Part: This lasts from 11:00 a.m. to 11:55 a.m., viz. 55 mins. The activities are: (a) assembly, 15 mins. (in the yard); (b) story-telling, 10 mins. (on the mat); (c) Bangla (vernacular), 10 mins. (on the mat); (d) mathematics, 10 mins. (on the mat); (e) news/ roll-call, 10 mins. (on the mat).

Second Part: The second part is from 11:55 a.m. to 12:55 p.m., viz. 60 mins. The following activities are performed : (a) Bangla reading, 2 mins.; (b) mathematics, 20 mins.; (c) assignment on language, 18 mins., (d) creative work, 20 mins.

Third Part: Duration 12:55 p.m. to 1:20 p.m., or 25 mins. Activities are (a) English, 10 mins. (board); (b) social studies, 10 mins. (on mat); (c) religion, 5 mins. (on mat).

Fourth Part: Duration 1:20 p.m. to 2:00 p.m., or 40 mins. Activities: (a) English reading, 5 mins.; (b) project on science, 15 mins.; (c) English, 10 mins.; (d) mathematics, 10 mins.

Weekly project

Learners of grades 4-5 do project work on a certain topic every week. Activities of the project can be divided into three types:

1. *Class book:* Learners prepare a class book on a particular topic or issue every week. This book contains a picture on the selected topic along with a few sentences beneath it to explain what the drawing represents. The learners draw a picture and then describe the guiding idea behind it.
2. *Personal book:* Besides preparing the class book, learners have to prepare individual books for themselves. They draw pictures for their own books. The learners who are independent writers put down their ideas beneath the pictures, while the teacher helps or writes on behalf of those who have not yet attained the skill of creative writing.
3. *Wall chart:* These are always hung at head level of the learners. In addition to the class book and the personal books, learners draw pictures on some object or topic on the wall In a month thirty pictures are drawn by the learners.

Learning materials

In GSS schools, children read between 12 and 24 books within the first two years. The teacher uses a variety of visual materials and games. Emphasis is placed on creative thinking and writing. For Bangla, children of GSS schools read a lot of books in comparison with their counterparts in government primary schools. In the GSS schools, the children in grade 1 read 8 books, in the grade 2, 14 books, and in grade 3, 14 books, while children in the government schools read only one book in each grade.

Special features of the GSS model

In the traditional system of primary education in government primary schools, rote learning is the normal practice and the children feel bored in the classroom. On the other hand, some of the special features of the GSS system are as follows:

(a) Well-being of learners is enquired after in the classroom.

(b) Learners give news about the events that happened around their houses.

(c) Group work is carried out.

(d) In roll-call, only the number of a pupil is mentioned, not the name.

(e) Children learn in groups.

(f) The teacher listens to every learner's reading.

(g) In creative writing each learner has to draw a picture and write a description of it.

(h) Reading record of learners is maintained.

(i) Learners practise arithmetic individually.

(j) Reading games are organised.

(k) There is a weekly project.

(l) Children draw pictures on the wall chart and in their books.

(m) Learners do co-curricular activities twice a week.

(n) The teacher gives concrete examples.

(o) In order to increase vocabulary, the learner uses a dictionary.

(p) Spelling is prohibited while a learner reads.

(q) One teacher teaches all subjects.

(r) The teacher goes around to the learners.

(s) Most of the time the teacher remains standing in the classroom.

Developing a creative writer

Achieving independence of thought among children and the ability to produce their own ideas in writing is one of the objectives of the GSS programme. Children attain the skill by putting their ideas and observations beneath pictures drawn by them. Thus GSS produces a large number of independent creative writers from among the children.

In this method, instead of being treated as part of the classroom the child is treated as an individual. The teacher pays attention to each child and individual learning takes place; each child proceeds according to her or his own ability and pace.

Community plays a vital role

As part of its involvement the community has to donate land. The community is also involved in the house-to-house survey of children. They take the programme staff around and explain why the survey is being conducted, and generally ensure willing compliance. Once a school is established, monthly parent-teacher meetings are the basis of the interaction between the school and the community. Discussions are held concerning any problems that may have arisen over the past month, especially with regard to children's attendance or teacher absenteeism. A nominal fee of Tk. 1:00 (US 2.5 cents) for admission and Tk. 1 per month for education is paid by all parents.

Cost effectiveness of the GSS model

Lack of resources is the main constraint in providing universal primary education in Bangladesh. As such, attention needs to be given to ensure cost effectiveness. The unit cost of the GSS primary education programme is about Tk.800 per annum. This is comparable to the unit cost of BRAC schools.

C. CMES: A TECHNOLOGY-BASED BASIC EDUCATION PROGRAMME

The CMES (Centre for Mass Education in Science) was founded by Dr M. Ibrahim and originated from the publication and distribution of a science periodical by him.

The first step towards the goal of CMES was establishing basic schools which were designed to provide non-formal primary education to disadvantaged children and youth along with integration of technology training and marketing. The basic school is a grass-roots school within the home environment of the target group. Its inner campus consists of a small bamboo-and-mud house or shed provided by the beneficiary families right within their homesteads.

Twenty basic schools are served by a Rural Technology Centre for Basic Schools (RTCBS), covering an area equivalent to a union (the smallest administrative unit of Bangladesh). There are four levels of education in the system: *ankur* (germinating), *bikash* (developing), *Agrosor I* (Advanced I) and *Agrosor II* (Advanced II), roughly equivalent to grades 1, 2, 3 and 4 respectively, of the formal system, and each takes about one year. The first two levels are available in the basic schools and the last two only in the RTCBS.

The *ankur* and *bikash* levels mainly deal with the attainment of literacy, numeracy and some life skills, while the *Agrosor I* and *II* in addition to basic education provide opportunities to learn one or more technological skills directly relevant to income generation.

The learner: The learners of the system are from those children who never went to school or dropped out. There is no age bar for enrolment. Thus this can be an option for children who are too old to enter or go back to primary school. Pupils may be enrolled in either *ankur* or *bikash* level as appropriate. In each of these levels there may be 20-30 pupils at a time.

Care is taken to enrol at least 50 per cent girl children.

The teacher: Each basic school has one teacher. The teacher is a young man or woman from the locality who has finished or is about to finish high school. In some cases very successful pupils who have passed from RTCBS within the system itself are trained as teachers.

The RTCBS has two teachers and five assistant teachers. The teachers are diploma holders in technical education while assistant teachers are skilled artisans who in most cases have had a reasonable schooling and have been practising their trades in the locality. Currently, there are 20 RTCBS centres with 400 basic schools enrolling 23,000 learners.

In addition to the basic education programmes, there is an adolescent girls' programme. The aim is to free the girls from social constraints and free their creative energies. CMES conceived the necessity of undertaking this programme in 1992 with the objectives of undoing the injustice done to the girls, helping them have a chance to exercise their basic rights as human beings, and develop properly towards an empowered womanhood.

Participants: The participants are distributed in basic schools and the RTCBS of the unit. There are 5-10 girls in a basic school and the number is 25-30 in the RTCBS. All the participants are organised into groups of 5, usually one group in a basic school, and 5-6 groups in the RTCBS. Participants' homes are in the vicinity of the basic schools.

Credit to participants: There is a credit scheme for the girls. For this scheme CMES entered into an agreement with the Grameen Bank, who provided funds to extend credit to the participants with the same rules and interest rates followed by the GB itself. The loans are managed by the programme teacher in consultation with the headmaster/assistant headmaster. The teachers undergo intensive training on credit management, organised and conducted by GB.

Skill training: Adolescents, both from RTCBS and basic schools, receive skill training in trades which have immediate job opportunities. The skill training chosen by the girls is mostly garment making, tie-and-dye, batik, poultry farming, sericulture, soap-making, candle-making and pottery. But the RTCBS also offers other trades such as carpentry, metalwork, welding, machine repairing, etc., which are traditionally considered the domain of the boys. Most of the training is on the job, while real-life services and production are actually marketed to the community.

Marketing of products: Products of the girls like candles, soap and poultry have market demand both in the school area and in distant urban and commercial centres. For the marketing of the products, a salesman is usually hired for a unit. The salesman maintains contact with local traders and shop owners and delivers the goods. To give an incentive to participants in various trades and the artisan teacher, the profit generated from the sale of products is distributed among them every two months.

Conducting the sessions: During these contact sessions further education and skill training for the participants are conducted and feedback is taken to assess the progress of each individual participant.

The special teacher and the assistant teacher conduct the contact sessions in the RTCBS and in the 20 basic schools in rotation. The routine is so arranged that there is such a session every week in a particular school. The assistant teachers, who are skilled artisans in most cases, have had a reasonable schooling and have practised their trades in the locality. They have expertise in various technologies relevant to the programme of the BS system. But from time to time they have to learn new techniques from various sources.

One of the major objectives of the programme is to put its participants in a leadership position in the community. Towards this end, participants are assigned to take the lead in installing a sanitary latrine, mobilising children to be vaccinated against six deadly childhood diseases, making provision for safe water, motivating villagers to make compost, providing nutrition advice to mothers, and so on and so forth. In carrying out the above activities, a participant pays house-to-house visits.

(a) *Determination of health and nutrition status:* The task is usually carried out by simple observation, taking body weight and body measurements to identify the state of nutrition. The measurement is followed by a discussion which focuses on preventive measures for some common diseases. In some cases patients are referred to nearby health centres. Distribution of simple remedies such as riboflavin tablets, iron tablets and oral saline, etc., takes place.

(b) *Installation of improved cookers:* Earthen cookers traditionally used by the Bangladeshi women are not energy-efficient and safe. They produce smoke and turn the kitchens into gas chambers, causing a health hazard to women. Moreover, 40-50 per cent of the fuel is wasted. To solve the problem, an improved cooker has been evolved by a state-run scientific research organisation. The adolescent girls carry out extension work to introduce the cooker to rural women by making the cooker themselves and also ensure its use.

(c) *Keeping the environment clean:* In Bangladesh refuse from kitchen, cowshed and poultry stocks is carelessly dumped near the household. This causes serious environmental problems. The girls persuade the housewives to dig a compost pit and put all waste inside it. Thus the environment is kept clean and the compost is used as fertiliser in the vegetable garden.

Service centre of CMES

The service centre, which is located in the capital city, functions as the headquarters of CMES. The basic school system is planned, developed, managed, monitored and evaluated from the service centre. It carries out research on appropriate technology for the basic school system. By organising training and preparing relevant materials and prototypes it transfers the findings of experimentation to the school system. It also innovates technology suitable for the villages. It designs and develops curricula, teaching methods, instructional materials, teaching aids and training courses for teachers and monitors their effectiveness.

Programme outcomes

Access to credit, literacy and exposure to technology have enabled many girls to become assets to their families instead of liabilities. Most of the girls have attained a considerable degree of skill in different programmes, which has helped them to start businesses with financial assistance from CMES. In many cases, the girls have assumed the responsibility for the whole family, even their elder brothers. Girls are also joining the formal system of education. Many adolescent girls on completion of the *bikash* stage have been admitted to the formal system in the third grade. Others are in the fifth grade on completion of *Agrosor* stages *I* and *II*.

Cultural development of the poor

Extreme poverty as well as conservatism often make life devoid of decency as well as aesthetic sense. The conservatives usually discourage children from reading novels, poetry, rhymes, etc. Children were also prevented from enjoying music, sports, dance, and so on and so forth. The adolescent girls' programme provided its participants with the opportunity to recite, sing, dance, read classic novels, etc., which were previously the monopoly of the middle and the rich classes.

Women's mobility has increased

The programme threw a challenge to the seclusion of women imposed by the conservatives. Now a growing girl sits in a tea stall serving the customers and doing book-keeping. Similarly, girls are found to move with the household merchandise across the villages, sitting in a market corner on the weekly market day, and doing similar other activities previously done only by boys.

Marriage delayed

Their association with the programme has changed the attitude of girls towards life and work. Now the girls plan their lives. The choice of a husband is no more the sole business of the parents. Almost all of them express the desire to be self-reliant before marriage, and delay marriage until 20 years of age.

From acquaintance to solidarity

RTCBS organises assemblies of adolescent girls every two months. On an average, over a hundred participants assemble at each. Before joining basic school they would not know each other, and it provides them with the opportunity to develop solidarity among themselves.

Thus, through the CMES, elementary education with life orientation has been made available at the doorstep of the learners. Instead of rote learning the participants reflect on themselves as human beings. As girls, they examine their relationship with boys and other male members of the society at family, household and community levels. What they learn, they try to apply in practical life.

In its basic school system, CMES found that in the first grade, enrolment and attendance of the girls were equal to those of the boys. Their performance was also better in the system's education and skills training programmes. But when they grow a little older everything changes with their sudden dropping out.

The programme has produced some visible results, which are reflected in the positive change of perception about women's life, greater mobility of the girls beyond the village, participation of girls in production and sale of market commodities, organising cultural functions, etc. Some of the activities have a direct bearing on the community. For example, the cheap sanitary latrines and the cheap washing soap produced by the girls and sold in the locality have a direct effect on the health and sanitation of the target group.

CMES organises workshops for adolescent girls and mothers both at the premises of Rural Technology Centres and basic schools. The purpose of these workshops is to begin and continue a dialogue with the families on all aspects of the programme. It provides an opportunity for the older generation to understand the problems through their own life experience and lend their support to the new generation's effort towards emancipation.

D. THE GOVERNMENT SATELLITE SCHOOL PROJECT

The Government of Bangladesh has undertaken several projects such as Food for Education, Compulsory Primary Education Programme, stipend for girl students, etc. to combat the problem of low coverage, low attendance and high drop-outs. The satellite school project is one such project designed to increase coverage and attendance and reduce drop-outs. The project was undertaken on an experimental basis in 1992 with the opening of 62 schools.

In a sense, a Satellite School (SS) is a feeder school for the normal primary school. The planners were convinced that bringing schools to the doorsteps of learners would increase enrolment. In the second year another 138 schools were opened, bringing the total to 200. Of these, 144 were set up in four administrative divisions, namely, Dhaka, Rajshahi, Khulna and Barisal, and the remaining 56 schools are in Chittagong division.

Programme description: Usually a locality with very low literacy rate, high population density and difficult access to school (particularly for girls) owing to natural barriers (canal, bush, field) is chosen for the satellite school. Children of the 6-7 years age group are enrolled in grades 1 and 2. In the second year, a low-cost two-room school is constructed. It is usually established one kilometre away from a primary school, which is called the mother school.

Teacher selection: Teachers must be female and employed as volunteers. They receive a modest monthly allowance of Tk.500. The induction of women in the project is expected to contribute towards the participation of women in development.

Mobilising community support: In the first year no expenditure is made on the construction of the schoolhouse. The house or space is provided by the local people to be used as a classroom temporarily. The classroom can be housed in a mosque, *maktab* (religious learning centre), veranda or in the drawing room of a private house.

If the requisite number of learners, cooperation from local people, space for a classroom, land for constructing a building, etc., are available in the first year, only then is a two-room building constructed in the second year. The establishment of a satellite school demands the following:

- Class of 1st year to be housed and conducted in a mosque, *madrasah*, veranda or drawing room.
- If the government decides to construct a two-room schoolhouse, the community must provide 15-25 decimals of land.
- The community and guardians must send all their children to the satellite school and ensure regular attendance.
- Participation of local people in managing satellite schools is essential.

In order to obtain community support and motivate the local people, a committee is formed with the following composition:

(a) Thana Executive Officer (TNO) — Chairman

(b) Thana Education Officer (TEO) — Member-Secretary

(c) Two teachers from local high school (preferably women) and H/M of a primary school, nominated by the TEO — Members

(d) One UP Chairman nominated by the TNO — Member

(e) One villager interested in education nominated by the TNO — Member

One female teacher is inducted in the first year, and on the fulfilment of conditions another is recruited in the second year. The teacher must have passed the SSC (Secondary School Certificate). A committee chaired by the Thana Executive Officer (TNO) usually selects the teacher through advertisement. The committee invites applications, scrutinises them and finally selects a volunteer teacher purely on a temporary basis. The committee makes its selection using the following criteria: (a) the candidate must be resident within one kilometre of the school; (b) she must have passed the SSC; (c) candidates with certificates in education will get preference; and (d) she must be 18 years old or above.

The selected teacher has to perform the following functions:

- She must conduct classes for 3 hours a day.
- She has to organise and conduct mothers'/guardians' meetings to motivate them twice a month. Proceedings of the meetings are to be sent to the Deputy Director, Satellite School Project, Director of Primary Education, Dhaka, through the Thana Education Officer on a regular b
- She has to arrange the enrolment of 6-7-year-old children within one kilometre radius of the school, the number of learners being at least 50.
- She must retain the enrolled children for two years and take the initiative to get the learners enrolled in the mother school in the 3rd grade.
- She has to conduct a survey in the school catchment area and display a list of school-age children on the wall of the school.
- Monthly monitoring report is to be prepared and passed on to the Deputy Director, Satellite School Project, through the TEO/ATEO.
- Total working days should be like in government primary schools.
- For no reason is the school to be kept closed. If the teacher is unable to attend class, the ATEO will take necessary steps to keep classes running.
- The teacher must attend the training courses organised by the Directorate of Primary Education.
- If the teacher is not found satisfactory, her services would be terminated.

Supervision of satellite schools: Teachers of the mother school provide overall assistance to the volunteer teacher of the satellite school (SS), while the ATEO inspects the school and supervises the teacher locally. Furthermore, TEO and DEO visit the school and send a report to the Deputy Director, Satellite School Project. An instructor nominated by the superintendent of the District PTI (Primary Training Institute) visits each SS once a month and send a report to the DD.

Training for the satellite school teachers: The Satellite School section of the Directorate of Primary Education has designed a training module for teachers which contains such topics as community participation, teaching-learning methods, child psychology, evaluation process, teaching aids, management, etc. The training course is expected to be attended by the District Primary Education Officer, TEO, ATEO, teacher of the mother school, PTI instructor and the volunteer teacher.

Management: The Deputy Director, Satellite Schools, under the Directorate of Primary Education based at Dhaka, is in overall charge of the project. He is assisted by an Assistant Director and a Research Officer. There is also a Satellite School Managing Committee headed by the local Union Parishad chairman.

The project aims to achieve some crucial objectives like drawing community support by way of securing space for classrooms and donation of land for schools, organising guardians' meeting by the volunteer teachers, total coverage, etc. Moreover, there are three committees: for the selection of a location for the school, to involve the community in teacher selection, and for school management.

The voluntary teacher bears a work load heavier than her counterparts in government primary schools. The pilot project is being implemented centrally. Its outcome would help small NGOs to design or redesign their own NFPE programmes.

There are some similarities between the non-formal primary education programmes of the NGOs and the Satellite School Project such as hiring female teachers from the community, providing them with short training, recruiting the teachers as volunteers for which they receive an allowance, block teaching, etc. The objective of solving children's access problem to school due to geographical barriers has largely been achieved. The other objective of drawing community support seems to have been achieved to some extent. It also has succeeded in recruiting village women for the teaching jobs.

Conclusion

It is thus obvious that the NGOs in Bangladesh have been playing a leading role in introducing innovative non-formal education programmes. There are some obvious limitations in the present non-formal education programmes. These relate mostly to the issue of replicability of the programmes. In most of the organisations there are no middle-level staff between the co-ordinator and the field supervisor. As such, the project head has to give considerable attention to field-level supervisors. Generally, the period of training is also too short to help a teacher internalise the whole teaching-learning process. The teaching method calls for a teacher with long experience and it is difficult for a newcomer to attain mastery over teaching skills.

Another problem is the lack of resource materials. Although the major NGOs generally have more resource materials than the formal system, the supply proves to be inadequate. A number of small NGOs have taken steps to introduce the newer innovative models such as those of BRAC and GSS in their schools, and for this they have sent their staff to the respective organisations to undergo training.

In the beginning, some organisations like the GSS faced difficulty in training teachers who would implement the relatively complex teaching method. Their method demands teachers of higher capability than the average teacher. The problem was solved in the following way:

- a systematic minute-by-minute time-table for the teachers
- a focused curriculum so that the number of tasks is restricted
- adequate provision of low-cost materials and resources to permit and support individual and group work
- on-the-job training for six months so that new teachers are carefully inducted into the role they have to play and into the philosophy that lies behind it

Overall, it may be said that the public sector agencies should benefit from successful NGO-introduced innovative programmes. NGOs, moreover, will need sustained government cooperation and support to make their efforts a greater success through to the end.

List of Contributors

Parthasarathi Banerjee	Scientist National Institute of Science, Technology and Development Studies (NISTADS) Dr. K S Krishnan Marg, Pusa New Delhi 110 012
Shakuntala Bapat	Director Indian Institute of Education Centre of Education and Development of Rural Women J.P. Naik Path, 128/2 Kothrud Pune 411 029
Ravi Chopra	Photojournalist, 13/88 Block C-4/B, Janakpuri, New Delhi 110 058
Shivashankar Dube	N.K. Bose Memorial Foundation B-8/9, Bara Gambhir Singh Gauriganj, Varanasi 221 001
K. Gopalan	"Chameli" Cherumukku Trichur 680 020

Suman Karandikar	Hon. Director, ASTRA Indian Institute of Education J.P. Naik Path, 128/2 Kothrud Pune 411 029
Nita Mathur	Research Fellow Indira Gandhi National Centre for the Arts Janpath New Delhi 110 001
Ram Lakhan Maurya	Secretary N.K. Bose Memorial Foundation B 8/9 Bara Gambhir Singh Gauriganj Varanasi 221 001
Chitra Naik	Member Planning Commission Yojana Bhavan, Parliament Street New Delhi 110 001
Neeru Nanda	Director Navodaya Vidyalaya Samiti A-39 Kailash Colony New Delhi 110 048
Gedong Bagoes Oka	Kompleks Dosen F.S. 2 Jl. Pb Sudirman Denpasar 80114 — Bali (Indonesia)
Dinanath Pathy	Secretary Lalit Kala Akademi Rabindra Bhavan New Delhi 110 001

D. Patnaik	Block A, 2nd floor Adarsh Colony Caranzalem Goa 403 002
Devi Prasad	52 Godavari Apartments Alakananda New Delhi 110 019
N. Radhakrishnan	'Sabarmati' Mitrapuram Paranthal P.O. Pathanamthitta Distt. 689 518/ Kerala
Baidyanath Saraswati	Unesco Professor Indira Gandhi National Centre for the Arts Janpath New Delhi 110 001
Haku Shah	16, Nemnath Society Narayan Nagar Road, Paldi Ahmedabad 380 007/ Gujrat
A.M. Sharafuddin	Director Campaign for Population Education 4/6 Block D, Lalmatia Dhaka 1207 (Bangladesh)
Pataraporn Sirikanchana	Hqrs of the World Fellowship of Buddhists 616 Sukhmvit Road (In the Benchasiri Park) Bangkok 10110 (Thailand)
Dwarko Sundrani	Director Samanvaya Ashram Bodhgaya 824 231/ Bihar

Kapila Vatsyayan
Academic Director
Indira Gandhi National Centre for the Arts
Janpath
New Delhi 110 001